T0074775

Textile Design

Textile Institute Professional Publications

Series Editor: The Textile Institute

Care and Maintenance of Textile Products Including Apparel and Protective Clothing
Rajkishore Nayak and Saminathan Ratnapandian

Radio Frequency Identification (RFID) Technology and Application in Fashion and Textile Supply Chain
Rajkishore Nayak

The Grammar of Pattern
Michael Hann

Standard Methods for Thermal Comfort Assessment of Clothing
Ivana Špelić, Alka Mihelić Bogdanić and Anica Hursa Sajatovic

Fibres to Smart Textiles: Advances in Manufacturing, Technologies, and Applications
Asis Patnaik and Sweta Patnaik

Flame Retardants for Textile Materials
Asim Kumar Roy Choudhury

Textile Design: Products and Processes
Michael Hann

For more information about this series, please visit: www.crcpress.com/Textile-Institute-Professional-Publications/book-series/TIPP

Textile Design

Products and Processes

Michael Hann

CRC Press
Taylor & Francis Group
Boca Raton London New York

CRC Press is an imprint of the
Taylor & Francis Group, an **informa** business

Cover Image: Detail of hand-woven saddle bag for camel. Early-twentieth century, Iran. Unknown designer. Item held at the University of Leeds.

First edition published 2021

by CRC Press
6000 Broken Sound Parkway NW, Suite 300, Boca Raton, FL 33487-2742

and by CRC Press
2 Park Square, Milton Park, Abingdon, Oxon, OX14 4RN

© 2021 Taylor & Francis Group, LLC

First edition published by CRC Press 2021

CRC Press is an imprint of Taylor & Francis Group, LLC

ISBN: 9780367313081 (hbk)
ISBN: 9780367313067 (pbk)
ISBN: 9780429316173 (ebk)

Typeset in Times LT Std
by Deanta Global Publishing Services, Chennai, India

Contents

Series preface

The aim of the *Textile Institute Professional Publications* is to provide support to textile professionals in their work and to help emerging professionals, such as final year or Master's students, by providing the information needed to gain a sound understanding of key and emerging topics relating to textile, clothing and footwear technology, textile chemistry, materials science, and engineering. The books are written by experienced authors with expertise in the topic and all texts are independently reviewed by textile professionals or textile academics.

The textile industry has a history of being both an innovator and an early adopter of a wide variety of technologies. There are textile businesses of some kind operating in all counties across the world. At any one time, there is an enormous breadth of sophistication in how such companies might function. In some places where the industry serves only its own local market, design, development, and production may continue to be based on traditional techniques, but companies that aspire to operate globally find themselves in an intensely competitive environment, some driven by the need to appeal to followers of fast-moving fashion, others by demands for high performance and unprecedented levels of reliability. Textile professionals working within such organisations are subjected to a continued pressing need to introduce new materials and technologies, not only to improve production efficiency and reduce costs, but also to enhance the attractiveness and performance of their existing products and to bring new products into being. As a consequence, textile academics and professionals find themselves having to continuously improve their understanding of a wide range of new materials and emerging technologies to keep pace with competitors.

The Textile Institute was formed in 1910 to provide professional support to textile practitioners and academics undertaking research and teaching in the field of textiles. The Institute quickly established itself as the professional body for textiles worldwide and now has individual and corporate members in over 80 countries. The Institute works to provide sources of reliable and up-to-date information to support textile professionals through its research journals, the *Journal of the Textile Institute*[1] and *Textile Progress*[2], definitive descriptions of textiles and their components through its online publication *Textile Terms and Definitions*[3], and contextual treatments of important topics within the field of textiles in the form of self-contained books such as the *Textile Institute Professional Publications*.

REFERENCES

1. http://www.tandfonline.com/action/journalInformation?show=aimsScope&journalCode=tjti20
2. http://www.tandfonline.com/action/journalInformation?show=aimsScope&journalCode=ttpr20
3. http://www.ttandd.org

Preface

From the dawn of known history there has been the desire to cover, protect and shelter the human body, initially using leaves, animal skins or other naturally sourced materials and, in many instances, replacing these with textile cloths created from fibres. It is not known whether independent discovery or the transfer of knowledge from elsewhere allowed different cultures to use fibres and to manipulate these to create continuously twisted structures known as yarns and, then, to take two sets of these and to interlace them at right angles to produce woven structures, achieved using a mechanism known as a loom. Equally, the scholarly world is unaware of the origins of the realisation that yarns could be manipulated to produce various looped or knotted structures, from which came into being the practices of knitting, netting, crochet and various related techniques. Likewise, the exact origins of felting are unknown, though, initially, this was simply a process which combined various animal fibres with scaly surfaces (particularly sheep's wool) through entanglement, encouraged by the presence of moisture, heat and friction.

The title of this book may seem ambitious, at least from the viewpoint of a knowledgeable textile technologist or textile scientist; it should be stressed that only a relatively small number of textile products is identified and only an outline of the stages of fibre processing is given within the covers of this single volume. It is believed, however, that the subject matter dealt with is sufficient in detail to be of value and to address the preliminary needs of textile and fashion designers as well as others requiring a rudimentary or introductory knowledge of the successive stages of textile processing, especially within an industrial context.

It is often the case that textbooks are expected to give such a level of detail that they can largely replace standard teacher-to-student interaction. This expectation cannot be realised here. Rather, this book can function as a reminder, an *aide mémoire*, and can support a traditional teaching role and not displace it. Nevertheless, the content of this book covers the main subject components of textile-technology courses taught to textile-design, fashion-design, fashion-communication, textile-marketing or textile-management students at colleges and schools worldwide and, as such, presents a brief outline of all the important processing areas and identifies the more important textile products. In terms of the space and attention devoted to each subject area, this has been determined partly by the level of subject knowledge of the author and, to some extent, by the perceived relative importance of the subject area to the intended principal audience as well as restrictions on space permitted by the publishers. Apologies are extended particularly to readers and colleagues whose expectations (in terms of the proportion of text and attention devoted to their area of expertise) have been dashed.

An emphasis is placed on the characteristics and properties of the more important fibrous raw materials and textile products, and a recognition is made of relevant processing techniques. In all instances, only the major types are presented, but should the reader require further knowledge or awareness beyond this, relevant additional sources are identified within the text.

All illustrative material is labelled numerically with respect to its intended chapter.

Acknowledgements

The author is indebted to I. S. Moxon for his characteristically constructive review, useful advice and debate. Significant contributions to the provision of illustrative material were made by Joanna Wilk (JW) and Jia Zhuang (JZ), as well as staff at the University of Leeds Library. Useful commentary on parts of the text was made by P. Henry, S. Russell, E. Gaston, K. C. Jackson, K. Wells, C. Cathcart, J. Winder and R. Blackburn, as well as staff and students at Donghua University (Shanghai). Thanks are due also to R. Murray, and others associated with the Textile Institute, as well as the editorial staff at Taylor and Francis. For past inspiration, I am indebted to J. Boomer, T. Ross, G. M. Thomson, W. Mackie, P. Grosberg, S. Hunter, M. Dobb, L. Peters, P. T. Speakman, P. Howarth, C. Hammond, C. S. Whewell, G. A. V. Leaf, I. Holme, J. W. Bell, M. Woodhouse, J. Jones, K. Blackburn, K. Hepworth, B. Hepworth, V. Whitehead, R. Grava, D. Lloyd, S. Harlock, J. A. Smith, K. P. Hann, M. and J. Large, E. and T. Hann, M. and B. O'Neill, R. and T. Mason, D. and L. O'Neill, J. Rosenthal, J. Hackney, B. Whitaker, J. MacMahon, P. Mc Gowan, M. Rossi, D. Holdcroft, A. Watson, P. Turnbull, A. Hollas, P. Byrne, P. Gaffikin, H. Coleman, C. Bier, R. Matteus-Berr, H. Diaper, D. Washburn, M. O'Kane, M. Anderson, C. Cini, D. W. Crowe, H. Mee, H. Hubbard, E. Broug, B. Thomas, D. Huylebrouck, H. Zhong, R. McTurk, J. Parker, C. Wang, S. Han, L. Bulsara and X. Lin. Last but not least, the author is pleased to thank N. B. Hann, E. A. Hann and H. C. Hann; without their unfailing support, this volume would not have been possible. The author accepts responsibility for all errors, omissions or false statements.

M. A. Hann
Leeds 2020

Author

Professor Michael Andrew Hann (BA, MPhil, PhD, FRAS, FRSA, CText., FTI) holds the Chair of Design Theory at the University of Leeds (UK). He is a graduate in textile design and worked for a brief spell in the Irish linen industry. He was the founding director of the University of Leeds International Textiles Archive (ULITA) and was a past Pro-dean for research with the Faculty of Performance, Visual Arts and Communications at the University of Leeds. He has published across a wide range of subject areas, has given numerous keynote addresses at international conferences and is a specialist in design geometry. He has interests that extend across various areas, including sustainable textile manufacture, diffusion of technological innovations, craft textile printing and symmetry in regularly repeating patterns. Recent publications include: *The Grammar of Pattern* (CRC Press, 2019) and *Patterns: Design and Composition* (co-authored with I. S. Moxon, Routledge, 2019). He has held adjunct, visiting and invited professorships at institutions in Brussels, Seoul, Hong Kong, Taichung (Taiwan) and Shanghai.

Reference sources for illustrative material

Figures were re-drawn, derived or developed from the following sources:

Bauman, J. (1987), *Study Guide to Central Asian Carpets*, Islamabad: Central Asian Society.

Bilgrami, N. (1990), *Sindh Jo Ajrak*, Karachi: Department of Culture and Tourism.

Hann, M. A. (2005), *Innovation in Linen Manufacture*, Textile Progress series (vol. 37, no. 3), Manchester: The Textile Institute.

Hann, M. A., and B. G. Thomas (2005), *Patterns of Culture. Decorative Weaving Techniques*, Ars Textrina series no. 36, Leeds: ULITA.

Humphries, M. (2004 [2000]), *Fabric Reference*, Upper Saddle River (NJ): Pearson Education.

Miles, L. W. C. (1994), *Textile Printing*, second edition, Bradford (UK): Society of Dyers and Colourists.

Murray, R. (1981), *The Essential Handbook of Weaving*, London: Bell and Hyman.

Sawbridge, M., and J. E. Ford (1987), *Textile Fibres under the Microscope*, Manchester: Shirley Institute.

Tubbs, M. C., and P.N. Daniels (1991), *Textile Terms and Definitions*, Manchester: The Textile Institute.

Website: Victoria and Albert Museum at https://www.vam.ac.uk/ (accessed 9.00 am, 10 March 2020)

1 Introduction

The purpose of this book is to identify the raw materials and the more important products and processes of relevance to the design and manufacture of textiles. The aim is to capture the essence of subject areas of importance to textile and fashion designers and others who are not so much concerned with the engineering and scientific aspects of the subject but rather need to develop a general awareness of fibres, yarns and cloths, their properties and how they are processed. Largely, the order of chapters follows the chronological stages of processing from fibrous raw materials to finished cloth and its testing.

Textile traditions are numerous worldwide, and it appears that most cultures developed their own unique means and approaches to the processing of fibres. All approaches cannot be identified and explained in this present book. Rather, the intention is to underline the possibility of great diversity in practice, and to suggest that invariably there will be numerous means of achieving one aim. In the industrial context, which is the principal focus here, the emphasis is on fewer rudimentary procedures, less labour intensity and a higher degree of mechanisation than was traditionally the case.

Chapter 2 identifies the various fibre types available to designers and highlights the properties and possible end uses of each. It should be recognised that often in popular usage the term 'wool' is used to refer to a wide range of fibres in addition to those sourced from domesticated sheep, especially mohair and cashmere. In the context of this book, however, the term 'specialty hair fibres' (a term found commonly in textile textbooks) is used instead to refer to all animal fibres mentioned here other than fibres sourced from domesticated sheep and silkworms. Meanwhile the term 'wool' is reserved only for fibres from domesticated sheep. Chapter 3 outlines the various means and stages of creating a twisted fibrous structure known as a yarn. Occasionally, the term 'thread' is found in various texts to refer to two or more yarns twisted together into one structure, and at other times, the term is used interchangeably with the term 'yarn'. At this stage, it is worth noting that in this present book the term 'yarn' will be used throughout, even where a twisted structure, such as a fancy yarn, for example, has more than one component part. Chapter 4 is focused on providing an explanation of weaving, the principal means of creating a textile through the interlacement of yarns, and chapter 5 on explaining knitting and other manipulation techniques of relevance to textiles and their embellishment. In all cases the term 'cloth' is employed, rather than fabric, to refer to the textile produced by either weaving or knitting, or any other technique. It is believed that the term 'fabric' should be reserved for use only when the cloth has been fully finished and is ready for immediate application to an anticipated end use. Chapter 6 presents an outline of the nature of felted and bark cloths and some other textiles often referred to collectively as nonwovens. Chapter 7 is concerned with colouration using dyestuffs; the main dyes and associated techniques are identified also. Chapter 8 identifies various block-printing and other techniques where dyestuffs are removed, resisted or encouraged to adhere to areas of the cloth or yarn. Chapter 9 outlines the nature of a selection of techniques used to finish textiles in order to make them perform better in their intended end use. Chapter 10 identifies a range of motifs and regular patterns. Chapter 11 considers aspects of the preparation of a collection of textile designs. Chapter 12 outlines various tests used in the appraisal of textiles and chapter 13 identifies relatively recent technological innovations and environmental concerns associated with textile manufacture. Concluding comments are presented in chapter 14.

The importance of scientific and technological enquiry in modern times to the discovery and development of fibre types and further understanding of known fibre types should be mentioned in this book, for without such enquiry (initially into natural fibre forms) an understanding of the requirements of manufactured fibres and how to process them would never have been achieved. What appears certain is that forming the basis of every major fibre discovery or processing breakthrough was extensive background enquiry. In the early- to mid-twentieth century, numerous educational institutions worldwide laid the basis among individuals for future discovery and invention. Formal research organisations played an important role, as did the large textile-machinery manufacturers and multinational fibre-producing companies together with various international professional associations. In the context of textiles, probably the most important technological developments in the late-twentieth and early-twenty-first centuries were the introduction of microfibres and nanofibres (both fine fibres with the latter even finer than the former), the introduction of high-performance fibres (including aramid fibres such as Du Pont's *Kevlar*) and the wide spread adoption of environmentally friendly fibres (such as lyocell). Also of importance was the use of known fibres in composites (larger structures where, often, fibres were included because of their favourable properties, especially their ratio of strength to weight), the introduction of autolevelling to ensure more regular yarns, the incorporation of electronic features into textiles, the massive expansion in applications for nonwoven varieties of textiles and developments relating to plasma finishing and digital printing. All of these had far-reaching implications for the textile industry in general. Of great importance also was the adoption of digital forms of control throughout the processing sequence, leading to less labour associated with textile manufacture and the automatic transfer between stages of manufacture.

Important industrial changes included the migration of textile and clothing manufacture away from centres in Europe to China and other parts of Asia, particularly in the wake of what became known as fast fashion. Meanwhile textile-manufacturing activities in the USA were largely retained. Major concerns, in the early-twenty-first century, globally, particularly in the more economically developed regions, were ethical manufacture (circumstances where the adult workforce was paid a reasonable wage, over a reasonable number of hours, in safe and clean working conditions), sustainability (a term used to indicate the use of raw materials not detrimental to the environment in the long term) and durability (where a product was designed to function fully into the long term). An appraisal of the nature of durability was given by Annis (2012) and a detailed review of associated issues as they related to textile manufacture was provided by Muthu (2017). By the end of the second decade of the twenty-first century, it was readily apparent that recycling (a term used to refer to the re-claiming and further processing of constituent fibres, yarns or cloths after their use for another purpose), upcycling (a term used to refer to the repair and probable re-design of a textile in order to extend its useful life) and ethical manufacture had captured the consciousness of the buying public. It was believed that recycling and upcycling could contribute positively to sustainability. Manufacturers and retailers were keen, therefore, to become associated with these developments as it was believed that this would enhance their profitability. There appeared to be a rush towards dissociation from any form of manufacture which could be deemed to be unethical, unsustainable or non-durable. Despite this, the environment continued to deteriorate and the overall contribution of textile processing and related manufacture to this deterioration seemed to have continued. Radical legislative approaches from governments were largely absent and a refocusing of the entire textile and related industries on the use of more sustainable means of manufacture had not occurred by the beginning of the third decade of the twenty-first century. At the time of manuscript submission, in early summer of 2020, the Covid-19 virus had advanced significantly worldwide, with substantial loss of life and the decimation of most national economies. It may be the case that between this submission and the time when the virus is finally dispelled, industrialists and policy makers will take time to reflect on how best to re-start their economies. In the textile context, for example, it may be worth considering and implementing some of the suggestions in publications such as *Circular Economy in the Textile Sector*, commissioned by the German Federal Ministry for Economic Co-operation and Development (BMZ) (Markschläger and Edele, 2019). This publication calls for fundamental changes in the way textile products are designed.

A brief explanation of the spelling conventions used in this book should be given. There is much variation throughout textile textbooks in the use of upper-case or lower-case letters, particularly with words such as 'Jacquard', 'Milano' and 'Roma'. In this present book, upper-case letters will preface both place names and personal names. Words such as Paisley and Jacquard will be spelt using upper-case P and J, respectively. In addition, because of the association of this book with the Textile Institute and its international headquarters in Manchester, British spellings (e.g. of terms such as 'fibre', 'labour' and 'colour') are maintained throughout.

REFERENCES

Annis, P. (ed.) (2012), *Understanding and Improving the Durability of Textiles*, Cambridge, UK: Woodhead.
Markschläger, F., and A. Edele (eds.) (2019), *Circular Economy in the Textile Sector*, Bonn: Deutsche Gesellschaft für Internationale Zusammenarbeit (GIZ) GmbH.
Muthu, S. (2017), *Sustainable Fibres and Textiles*, Cambridge, UK: Woodhead.

2 Textile fibres

2.1 INTRODUCTION

Fibres used in textile manufacture can be classified as natural or manufactured fibres. Further sub-categories are also common. Natural fibres are of vegetable, animal or mineral origin. All vegetable fibres are considered cellulosic (although each vegetable fibre type holds a differing proportion of cellulose). Among these are bast fibres (such as flax, or other fibres obtained from within the stems of plants or trees), leaf fibres (such as sisal) and seed fibres (such as cotton). The characteristics of each of the three types of vegetable fibre are outlined in section 2.2.

Numerous natural sources provide fibrous matter of one kind or another. Fibres used in the manufacture of textiles require a range of structural properties. They need to be like flexible rods, without branches and with a significantly longer length than the cross-sectional width. Textile fibres (referred to simply as fibres in the subsequent sections) may be classed by reference to their origin; some, like those mentioned in the paragraph above, are naturally occurring and are known as natural fibres, some are reconstituted from natural sources and are known as regenerated fibres and some, manufactured from assemblies of chemicals, are known mainly as synthetic fibres. The term 'manufactured fibres' (rather than man-made fibres) will be used in this present book to refer collectively to regenerated and synthetic fibres.

No matter which application may be envisaged, raw-material selections are critical as these must not only have the desired properties but also be suited to available means of processing. Considering that raw-material costs are a substantial proportion of the total costs of production, the necessity to ensure that suitable fibrous materials are selected seems paramount. For those working in textiles, be it in manufacture, design, display or analysis, constituent fibres are a fundamental aspect. Although underlying arrangements of fibres and cloth construction, as well as applied finishes, are of crucial importance, it is the constituent fibre type which will have the paramount influence on the performance of the finished textile. Therefore, it is necessary often to identify constituent fibre content of cloths, maybe through microscopic examination or by burning, staining or solubility tests.

In the spinning of manufactured fibres in continuous-filament form one of several general systems has been used. Basically, fibres may be obtained through melting or dissolving constituents in a solution, with the former known as melt spinning and the latter, occasionally, as wet spinning (though there are various further classifications possible) (Figures 2.1 and 2.2). In each case, the fluid is extruded through a thimble-like device (with numerous holes), known as a spinneret. Once filaments have been formed, these are stretched or drawn out to improve their inherent physical properties. Resultant fibres may also be texturised using heat to impart coils or crimps along their length, thus increasing their bulk. The process can influence

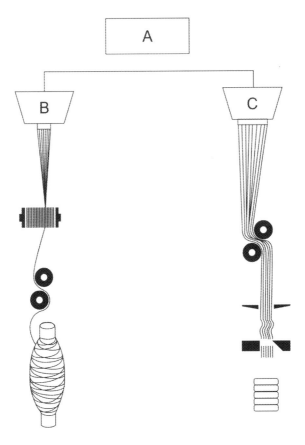

FIGURE 2.1 Melt spinning (A) with either continuous-filament processing (B) or staple-fibre processing (C). Re-drawn by JW.

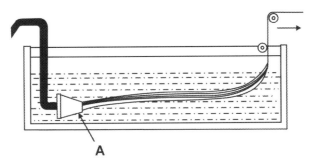

FIGURE 2.2 Wet spinning (viscose), A = spinneret. Re-drawn by JW.

various performance characteristics of resultant yarns and cloths with improvements to degrees of elasticity and handle, as well as heat retention. Hearle, Hollick and Wilson (2001) provided a substantial review of yarn texturising technologies and products.

Commonly, with fibres such as lyocell (where a technique known as solvent spinning is used), the bulk of solvent is re-claimed and returned to the process. All categories of manufactured fibres can undergo further processing in continuous-filament form or, alternatively, can be chopped into staple form prior to further processing.

The objectives of this present chapter are therefore to identify and discuss briefly the most common natural fibres (from both plant and animal sources) and manufactured fibres (as noted previously, this is a collective term used here to refer to regenerated fibres and synthetic fibres). It should also be mentioned that various other fibres can be identified, including asbestos, a naturally occurring mineral fibre that fell out of popularity during the late-twentieth century, due largely to major health concerns; as asbestos has fallen out of use, it is not dealt with further in this chapter. However, it does seem appropriate to mention a further three fibre types (glass, carbon and metallic fibres), as each has played an important role in early-twenty-first-century textile technology; each of these fibre types will be discussed briefly in this chapter alongside various synthetic fibres. Fibres under the general heading of ceramic fibres, although not mentioned further here, are an additional possibility; these were used in various high-performance areas, especially where fire proofing was a required feature. The most renowned twentieth-century publications concerned with fibres were those produced by Cook (1993a and b [1984a and b]), with volume 1 concerned with natural fibres and volume 2 with manufactured fibres.

2.2   PLANT FIBRES

The most common plant fibre is cotton. The bulk of archaeological evidence suggests its early use in South Asia (particularly India and Pakistan), with numerous cotton-cloth types developing there. According to Barber, cotton has been used as a textile fibre for around 5,000 years (Barber, 1991: 32–33). In the British context, the city of Manchester and surrounding areas in the north-west of England, during and immediately after the period known as the Industrial Revolution (of the eighteenth and nineteenth centuries), became associated closely with cotton processing. Good reviews of the historical context were provided by Beckert (2014). By the early-twenty-first century, cotton was cultivated in numerous countries worldwide, with the largest quantities grown in the People's Republic of China, Brazil, India, Pakistan, the USA and Uzbekistan; collectively these accounted for around four fifths of global cotton harvesting as well as ultimate processing and manufacture.

Cotton fibres grow from the seedpod (or boll) of the cotton plant, with length measurement of fibres (or staple length) between 0.5 and 1.5 inches (or 1.27–3.81 centimetres) depending on the species. Fibres of longer staple length are of higher quality. It is a versatile fibre, combining softness and strength with economy of extraction and relative ease of processing. Most cotton is off-white and needs to undergo bleaching before dyeing. The fibres may be handpicked or harvested by machine and are separated partly from the seeds and other non-fibrous material using a machine known as a gin. Numerous fabric types, using 100 per cent cotton, can be identified, including calico, towelling of various kinds, denim and muslin with applications across fashion apparel and interior end uses. Cotton yarns may be twisted together and used as sewing threads, in embroidery and lacemaking, and cotton's by-products can be put to further use, with the very shortest fibres destined to stuff mattresses and upholstery, and the seeds from the ginning process converted into animal feed or soil fertiliser. However, there are substantial drawbacks. Cotton is the most unsustainable, and least environmentally friendly, commonly used fibre; when produced by conventional means, it requires vast quantities of water, pesticides and fertilisers when growing, as well as further quantities of both water and chemicals during processing. Late-twentieth- and early-twenty-first-century attempts to remedy this included genetically modified (GM) and organic types of cotton, both aimed at reducing the need for pesticides and fertilisers. A well-focused review of biological information of relevance to genetically modified cotton was produced by the Australian government (2008) in the early-twenty-first century; over a decade later many of the issues raised were still of relevance. Also, by the beginning of the third decade of the twenty-first century, cotton continued to require large quantities of water both for its growth and for its processing. Because of this, the focus of attention with a relatively small number of processors worldwide was on the use of unbleached and undyed cotton which required less water in processing. Cotton fibres consist predominantly of cellulose and, when viewed longitudinally under a strong microscope, appear to have flat, twisted, ribbon-like structures. When viewed in cross-section, each of the fibres is kidney shaped, with thick walls and a small lumen (or central core). Generally, untreated cotton fibres lack lustre, are relatively strong when dry and even stronger (up to around one quarter increase) when wet. The fibres are not extensible, crease badly, are good conductors of heat and are thus suited for summer wear. However, cloths created from 100 per cent cotton fibres soil readily, shrink on washing, are weakened by concentrated alkalis and strong light, and are damaged badly by acids and mildew. They have poor resistance to wear when compared to certain synthetics (such as polyester, for example). So-called easy-care cotton shirts and blouses wear badly on collars and cuffs. Cotton fibres dye well with good colourfastness, using a wide variety of dye classes. Cotton fibres are used extensively in blends, particularly with polyester across a wide range of end uses. Probably the finest variety is Sea-Island cotton, which is among the most expensive also. A well-focused review, considering several important twenty-first-century issues, was provided by the ICTSD (2013), an organisation concerned with sustainable development. An extensive review of developments in the knowledge of cotton and its processing was provided by Gordon and Hsieh (2006).

Coir fibre is extracted from coconut husks. When sourced from fully ripened coconuts, a brown-coloured fibre results and, when taken from yet-to-ripen coconuts, a white or pale-brown fibre is the outcome. The brown-coloured fibre type is exceedingly coarse, not too flexible, though stronger than cotton and resilient to abrasion; it is used predominantly in hard-wearing brushes. Meanwhile, the white or pale-brown coir fibre type is softer (and less strong) and may be twisted into rope or woven into floor mats. Fibres have good resistance to both microbial attack and salt water. Important producers include India and Sri Lanka as well as Brazil, Indonesia, Vietnam, Thailand and the Philippines, where, typically, in each case, the fibres are harvested from coconuts grown by small-scale farmers and subsequently extracted by local mills. Other types of plant fibre include those found within stems or leaves. The former is known as bast fibre and the latter simply as leaf fibre. Typical bast fibres include flax, jute, hemp, ramie, kenaf, nettle and banana, and leaf fibres include sisal, abaca (or Manila hemp) and pina. Each is considered further below.

Bast fibres are distributed longitudinally within the stems of various plants and, to be used, need first to be separated from the woody matter that makes up the stem. With a few minor exceptions, different bast fibres undergo similar processing and associated terminology is broadly similar. Flax fibre is classified as a bast fibre and is obtained from within the stalk of an annual plant (identified in textbooks as *Linum usitatissimum*) which grows readily in temperate and sub-tropical regions. Archaeological evidence suggests that flax was probably the first fibre to be processed into a textile. According to Barber, flax fibres have been in use for around 7,000 years (Barber, 1991: 11). The fibre is associated closely with ancient Egypt, and its use is recorded in parts of Europe several centuries before the Common Era. The processing of the fibre in Britain and Ireland was stimulated by the arrival of Huguenot textile workers fleeing religious persecution in France during the late-seventeenth century. Particularly fine linen manufacture became associated with the city of Belfast and surrounding locations in the northern counties of Ireland, during the nineteenth and for much of the twentieth century. By the early-twenty-first century, a substantial proportion of world acreage devoted to the growth of the fibre was in the People's Republic of China and the Russian Federation. After the flax stalks are pulled (rather than cut, as a large percentage of the fibre is held underground), several processes are necessary in order to gain access to the fibres. First, stalks should undergo a rotting process, known as retting, traditionally carried out in dams, tanks or streams, or through a process known as dew retting (where the flax stalks are simply spread on the ground in fields and are thus subjected to prevailing weather conditions). Various forms of chemical retting, including the use of enzymes, were the subject of experiment in the early-twenty-first century. The lustre and colour of the final fibres are determined to a large degree by the system of retting selected. After the loosening of the fibres through retting, processes known as breaking (where the stalks are simply broken between fluted metal rollers), scutching and hackling help to separate the fibres further from unwanted woody matter and make the fibres ready for various stages of spinning. As processing continues, two types of fibre emerge: long fibres (known as line), which produce higher-quality yarns, and short fibres (known as tow), which produce yarns considered to be of lower quality. Traditionally, coarser yarns were obtained through dry spinning and finer yarns through wet spinning (with the fibres being passed through hot water, to soften gummy matter, prior to twisting). When fibre is separated from the woody matter by purely mechanical means, the resultant fibres are referred to as green flax. When viewed longitudinally under the microscope, flax fibres appear smooth and bamboo-like, with a narrow lumen, occasional cross markings or nodes and with few (if any) lengthways striations. In cross-section, the fibres have a sharp polygonal shape with five or six straight sides. Flax fibres consist of around 75 per cent cellulose, with the remaining quarter consisting mainly of pectin and various waxes. The fibres are more lustrous and stronger than cotton fibres and, like cotton fibres, show an increase in strength when wet. Flax fibres are around 20 per cent stronger when wet, compared to their dry form, but suffer a gradual loss of strength when exposed to sunlight for prolonged periods. Flax fibres are not elastic and, in cloth form, crease easily. Water is absorbed well and dries quickly, and good heat conduction makes cloth from flax fibres cool and thus suited to summer wear. Flax fibres do not soil easily due to the relatively smooth fibre surface. Hot bleaches and acids easily damage the fibres, though they are resistant to alkalis. Compared to cotton fibres, flax fibres have a poorer affinity to dyes and are more expensive to purchase. Relevant explanatory literature was reviewed by Hann (2005), while Sharma and Van Sumere (1992) provided a good explanation of processes also of relevance in the twenty-first century. Franck (2005) provided a comprehensive review of flax and other bast fibres as well as some leaf fibres. It is apparent that many of the processes associated with the processing of flax fibres are of relevance also to other bast fibres, such as jute, hemp, ramie, kenaf, nettle and banana. Each of these fibre types is reviewed briefly below,

Jute, referred to occasionally as the golden fibre (due to its yellow to pale-brown colour) is the cheapest bast fibre, and is naturally biodegradable with advantages of good dimensional stability, moderate strength, thermal conductivity, coolness and favourable ventilation properties. However, jute fibres are not as strong or as durable as flax fibres and show great variation in physical characteristics from fibre to fibre. The fibre can absorb close to one quarter of its own weight in water, though under humid conditions the fibre deteriorates. With the passage of time, the fibres show loss of strength and are sensitive to chemical attack but show good resistance to rot. The fibres are soft with a silky lustre and longitudinally, under a strong microscope, appear cylindrical in shape and, in terms of cross-section, have rounded, polygonal shapes each with a centrally located broad lumen (which varies greatly in dimensions, even along the same fibre). The fibres have high insulating and antistatic properties, and moderate moisture regain and low thermal conductivity. In the British context, the Scottish city of Dundee became associated closely with jute-fibre processing in the late-nineteenth century and for much of the first half of the twentieth century. By the early-twenty-first century, the bulk of both harvesting and processing was in Bangladesh and West Bengal (in India), with smaller amounts produced also in Nepal. In all cases, production levels were very sensitive to weather conditions. After cutting off the stalks (using an implement like a sickle), retting is carried out to help separate fibres from stems, with the stalks being soaked simply in a slow-flowing stream. After rinsing in clean water, the fibres are stripped from the stem manually, dried in the sun for a few days and compressed into bales in readiness for spinning. Fibres are used mainly in upholstery, carpets and rugs and, traditionally, in twines and as backing for linoleum. The introduction of various manufactured fibres (particularly polypropylene) in the late-twentieth century offered severe competition in traditional end-use areas. Because jute fibres are biodegradable, flexible, absorb moisture and drain well, they are well suited to various uses as geotextiles to prevent soil erosion and landslides.

Traditionally, after dew or water retting, stems of hemp (another bast fibre) went through breaking and scutching processes, with the fibres separating from stems more readily than with flax fibres. Green hemp (where fibres were separated from stems using mechanical means only, similar to arrangements for green flax mentioned earlier in this section) was a common form of production in the late-twentieth century. Hemp fibres are strong and durable but are coarser and stiffer than flax fibres. The fibres are fast growing without the assistance of pesticides or insecticides. When viewed longitudinally under a strong microscope, hemp is smooth and cylindrical with cross-marking nodes and a broad lumen and, in cross-section, the fibres appear partly polygonal. Fibres are long, strong and durable, dye well, are resistant to mildew, block ultraviolet radiation and have natural anti-bacterial properties. By the early-twenty-first century the top producing region in terms of fibre quantity was the People's Republic of China. Smaller amounts were produced also in parts of Europe, Chile and the Democratic People's Republic of Korea. Production was restricted in some countries where there was confusion between the fibrous variety of the plant and a highly potent narcotic variety. By the early-twenty-first century, in the People's Republic of China, hemp was often de-gummed and processed on flax or cotton machinery. Also, the fibres were used commonly to reinforce moulded thermoplastics in the automobile industry worldwide. Traditionally, common uses for processed hemp fibres included ropes, canvas and paper.

Ramie, another bast fibre, is also known as China grass. Ramie fibres are indigenous to Asia, and the main area of harvesting in the early-twenty-first century was the People's Republic of China, with relatively small amounts also in Taiwan, Japan, the Philippines and Brazil. The bulk of fibre was for home consumption, though small quantities were exported to Germany, Japan, France and the UK. Traditionally, the fibres were separated from stems through a process comprising initial soaking in water followed by simply peeling or scraping the fibres from the inner bark of the stem. Removal of fibres from stems by decortication (a mechanical, scraping process) was the norm and this was best practised when the fibres were fresh (otherwise, extraction was difficult). By the early-twenty-first century, mechanised means for separating fibres from stems had been developed and were in common use. Large quantities of gum needed to be removed from the fibre strands prior to spinning. Often gums were removed by repeated soaking and scraping. Soaking fibres in caustic soda for a few hours prior to scraping proved helpful, as did light bleaching and steeping in weak acid. After separation from gum, the fibres were rinsed, oiled and dried. Ramie fibres are exceedingly strong, especially when wet; they retain shape when in use and are wrinkle resistant. They are commonly blended with other fibres such as cotton, flax, silk or wool, which also improves their handle. Ramie fibres, when viewed longitudinally under a powerful microscope, are broad and irregular with cross markings distributed irregularly. When viewed in cross-section, ramie fibres appear oblong. Although stiff, with very low elasticity, and coarse and hairy to touch, ramie fibres are as strong and durable as flax fibres. They are invariably white and lustrous, absorb water readily, will launder easily and dry quickly. Although much of the natural lustre is diminished during weaving, this characteristic can be remedied through soaking of the yarns in caustic soda, rinsing and drying prior to weaving.

The kenaf plant has a long history of cultivation and processing, and the fibre, when processed, is visually like jute but is harder and stronger. In the early-twenty-first century, the main areas of agricultural growth were the People's Republic of China and India. The plant requires a minimal application of pesticides and fertilisers but needs to be retted to extract the fibres. The fibres are strong when compared to other natural fibres, and end uses traditionally have included twines, cordage and ropes. Although the fibres are shorter and coarser than jute, kenaf often substitutes for jute in sacking and bags. The plant is exceedingly quick-growing and the yield per acre is very high.

Nettle fibres are another form of bast fibres, extracted simply from the common stinging nettle and grown readily in most temperate zones. The fibres held within the stems are exceedingly strong, are soft, smooth and white, and can be used to produce a fine linen-like cloth. Seen under a strong microscope, the cross-sections are oval and, longitudinally, thick walls, irregular in width, are typical. After harvesting, the stems are retted and then boiled to help separate the fibres from the inner stem. Once separated, the fibres are hackled (a sort of combing process used also for flax fibres). The plant grows readily in disturbed damp ground. The fibres are related to flax and hemp and their growth and extraction offers great potential as a valid environmentally sustainable alternative to highly polluting cotton.

Banana fibre, also known as musa, is a strong natural fibre extracted from within the bark of the banana tree grown for its fruit. Thick, sturdy fibres can be extracted from the outer areas and softer fibres from the inner areas. The fibres can be used in the manufacture of ropes and woven floor mats as well as in various fringe areas such as tea bags, currency notes and as components of car tyres. The fibres are regarded as soft, yet durable and easily spun with high tensile strength and good extensibility; they also have good water and fire resistance.

Commercially important leaf fibres include sisal, abaca and pina. Each is described briefly below. Sisal is obtained from the leaf of the sisal plant, which matures between three and five years after planting and yields fibres for seven to eight years thereafter. Initially, in the first year of harvesting, the plant should yield around sixty leaves, then thirty leaves annually. Fibres, which are creamy-white in colour, are located longitudinally in the leaves. By the end of the second decade of the twenty-first century, sisal was grown extensively in Brazil, Kenya, Tanzania, Madagascar, Mexico, Haiti and the People's Republic of China. After harvesting of the leaves, fibres are extracted through decortication, where leaves are crushed between rollers and non-fibrous matter is scraped off; in much of Africa this is assisted by a water-based technique (which helps to remove waste matter) but elsewhere it is a dry, purely mechanical, process. Fibres are graded based on length and colour, and baled. Lower-grade fibres are destined for the paper industry, middle grades for cordage, ropes and twines, and higher grades for the carpet industry. Overall, sisal fibres are not suited for garment use. Competition from polypropylene weakened demand significantly in the late-twentieth century. Further end uses, in the early-twenty-first century, included specialty papers, carpets, wall coverings, mattresses and geotextiles. Sisal fibres are strong and durable, though stiff and rather inflexible. They resist deterioration by seawater and will accept a wide range of dyes. Seen under a strong microscope, longitudinally sisal fibres are cylindrical and have tapering ends and, in cross-section, show a central lumen with dimensions which vary greatly from fibre to fibre.

Also known as Manila hemp, abaca fibres are from the leaves of the abaca plant, native to the Philippines and from the same family as various banana plants. Plants grow well in tropical climates, particularly where volcanic activity has been a feature. When ready for harvesting, each leaf stalk is cut into strips and the fibres are removed by mechanical scraping (by hand). After washing and drying, the fibres are ready for processing. Abaca fibres have excellent mechanical strength, are lustrous, beige in colour and resistant to saltwater; traditionally, the fibres were processed for use as ships' rigging and ropes, as well as twines, fishing lines and fishing nets. In addition, the fibres were pulped to make sturdy Manila envelopes. The better fibre grades are fine and lustrous, with a light-beige natural colouring. By the early-twenty-first century, the world's largest-producing region was the Philippines. Although manufactured fibres had been substituted for abaca in many end-use areas, niche markets had developed for abaca clothing and furnishings. Further niche areas, in the early-twenty-first century, included currency notes (e.g. Japan), cigarette filter papers and tea bags, as well as composite additions in automobile parts, substituting often for glass fibre (which consumes much more energy in its manufacture). The plant has various favourable environmental features and its planting can assist greatly in minimising the effects of erosion in coastal areas.

Pina fibres are extracted from the leaves of the pineapple plant, which grows commonly in the Philippines (the main producer of pina fibres by the early-twenty-first century). Fibres have favourable mechanical characteristics, so are suited as additions to reinforcement composites (where the fibres impart improved structural strength). Traditionally, the fibres were used in various forms of formal wear, as well as in table linens and as clothing items destined for export.

Other cellulosic fibres include kapok and bamboo. Kapok fibre, extracted from the seed pod of a tropical tree called the 'ceiba', though soft and silky, is unsuited for conventional textile uses because the fibre is brittle and slippery (so is not easily processed); but it has been used as stuffing in upholstery and bedding, and in thermal and sound insulation. Bamboo, on the other hand, offers great potential for further exploitation as a sustainable textile fibre. It is a fast-growing grass, ready for harvesting after around four years, with no need for replanting because of the extensive root system. In the early-twenty-first century, the plant was grown extensively throughout Asia, particularly in the People's Republic of China. Once harvested (through cutting), vigorous re-growth takes place. In addition, the plant does not require pesticides or insecticides to grow and survives in climatic extremes, including high temperatures, droughts and floods. Fibres can be extracted via retting and processing arrangements can be like those developed for flax fibres. Although substantially more research needs to be conducted to confirm the suitability of bamboo fibres as viable alternatives to more polluting fibres, it appears that they offer great environmental credentials, including cheapness, low land usage, ease of cultivation and a fast growth rate, as well as a potentially wide range of applications.

Comprehensive explanation of the characteristics and processing of a selection of four fibre types (sisal, jute, abaca and coir) was provided by the Food and Agricultural Organisation of the United Nations (http://www.fao.org/economic/futurefibres/home

/en/). The FAO offered a focus on these four fibres as it was believed that these were ideal for sustainable development and future expansion in their manufacture.

2.3 ANIMAL FIBRES

Although, by the early-twenty-first century, animal fibres made up a relatively small proportion (probably no more than 5 per cent) of all fibres used globally, they continued to hold an important position (probably because of the range of unique characteristics offered by them). By the early-twenty-first century, fibres from sheep were by far the most-processed animal fibres. Animal fibres are protein fibres and are therefore related in their constituent chemical structures, but they differ remarkably from each other in terms of their precise physical properties.

Animal fibres of various kinds have been used historically. The wool (or hair) from certain mammals, including sheep, goats of various kinds, angora rabbits, llamas, alpacas, vicuñas, camels, yaks, buffalos and musk ox, is the most important. Meanwhile, filament extruded from the silkworm is the most common fibre of insect origin. As noted previously (in chapter 1), the word 'wool' is used in this book to refer to the fibres associated with sheep and not fibres from the coats of mammals other than sheep. Meanwhile, the term 'specialty hair fibres' is used to refer to fibres from mammals other than sheep, and the term 'kemp' reserved to refer to the outer fibrous coat (often of very coarse fibres covering the useful finer fibres beneath). With each animal, the quality of fibres obtained will depend to a large degree on the quality of nutrition, levels of care and temperature/climate experienced by the animals.

The domesticated sheep produces a fleece of soft, curly fibres used extensively in textile manufacture. Under a powerful microscope, sheep's wool fibres, when viewed longitudinally, appear to have a surface of overlapping scales. Generally, the fibres are crimped rather than straight, with increases in the number of crimps per unit length correlating with the perceived increases in fibre quality. For example, merino wool (a type perceived of the highest quality) appears to have the highest number of crimps per unit length. Lower-quality wool fibres have fewer scales per unit length and, for this reason, are more lustrous than higher-quality wool fibres. The fibres are very extensible and are crease resistant. When wet, the fibres lose much strength. They are bad heat conductors so, in cloth form, are warm and suited to colder climates. Wool fibres are highly absorbent and can hold up to one third of their own weight in water without feeling damp, but, when wet, the fibres do not dry quickly. Wool cloth does not soil easily, though it shrinks and felts without difficulty, is readily damaged by chlorine bleaches, can be attacked by insects and is damaged by mildew. The fibres are non-flammable, resistant to weak acids and dye well with a wide variety of dye classes, but are relatively expensive compared to cotton fibres. A review of advanced knowledge of wool fibre and its processing was provided by Simpson and Crawshaw (2002) and, subsequently, by Johnson and Russell (2008).

Cashmere fibres are obtained from the Kashmir goat, native to the Himalayan region, particularly Tibet and northern India. A fine undercoat of fibres is collected often by combing, and occasionally by shearing, the animals during the moulting season early each year. After collection, fibres are sorted and cleaned, with the coarser outer hairs (or kemp) being removed. Remaining fibres have natural crimp, and are suited to the production of fine, light-weight cloths. Due to a series of tiny air pockets, cashmere items are warm yet light-weight and, with low-relief scales on the surface, the fibres are lustrous and smooth. Natural colours vary from white to various greys and browns. By the early-twenty-first century, the fibres were regarded as luxurious, as well as rare and expensive. Leading global producers were based in the People's Republic of China, followed by the State of Mongolia, with small levels of production also in Australia, Iran, Pakistan, India, Turkey, New Zealand and the USA. The fibres were used commonly in knitted sweaters and in baby-wear items due, especially, to the warmth and softness provided. A variety of the fibres, known as pashmina, was regarded as the finest form of cashmere fibre and was used mainly in scarves and shawls produced often in the Kashmir region of north-west India.

Mohair fibres are from the fleece of the angora goat, which is shorn twice per year. Surface scales are thin, which make fibres smooth to touch and gives them their characteristic lustre. The fibres are moisture absorbing, flame and crease resistant, have good dye uptake and are used commonly in blends with other fibre types to enhance the overall quality. Tensile strength is higher than for top-quality sheep's wool such as merino. By the early-twenty-first century, the main producers were based in South Africa (in the Eastern Cape region, particularly), as well as the USA (mainly Texas). End uses included knitting and crocheting yarns and, in cloth form, stoles, scarves and high-quality blankets. Generally, finer fibres from younger animals were used for clothing, whereas coarser fibres from older animals were used in carpets.

Fibres from the alpaca (a domesticated camel-like animal native to South America) are fine, soft, warm and light-weight, with brightness and crimp, and are particularly suited to knitted and crochet end uses. The fibres are stronger than wool fibres from domesticated sheep and, by the early-twenty-first century, the principal end uses were as knitwear of various kinds; but fibres were also spun into yarns suited for weaving (mainly in the form of stoles and shawls). Partly hollow fibres account for alpaca's heat retention and insulation properties. Alpaca fibres blend well with wool fibres, as well as with mohair or silk fibres. By the early-twenty-first century it appeared that there was little scope for expansion of production in much of South America due to lack of grazing lands. Herds in North America and Australia were, however, expanding to the extent that each was destined to become a major world supplier.

Fibres are also sourced from the angora rabbit (not to be confused with the angora goat). The animal provides silky textured fibres, which are both warm and light-weight. Fibres are exceedingly fine and very soft to touch (probably due partly to the low relief of the fibre's scales). The hollow fibres allow for substantial heat retention, water absorption and easy dyeing. Low elasticity and low resistance to abrasion are further features. Various natural colours are available, including tan, various greys, mid-brown and black. Yarns of 100 per cent angora-rabbit fibre are occasionally used as effect yarns. Commonly, however, the fibres are blended with a fine sheep's wool, such as merino, or with lamb's wool. The world's main producing region up to around the middle of the twentieth century was France but, by the early-twenty-first century, this position was taken by the People's Republic of China (which produced around 90 per cent of the global total, involving 50 million rabbits). Smaller quantities of the fibres were produced in Argentina, Chile, the Czech Republic and Hungary. Further details of the fibre and its usage were provided by Dirgar and Oral (2014).

The two-humped Bactrian camel, native to the steppes of Eastern and Central Asia, is also a common source of fibres. The fibres were used traditionally by nomadic peoples as raw materials in the production of warm clothing, carpets and yurts. Exceedingly soft, lustrous, warm and fine fibres offer a good range of natural colours. By the early-twenty-first century, significant producers included the State of Mongolia, the People's Republic of China, Afghanistan and Iran. Fibres were highly regarded and added *caché* when blended with other fibre types. Often luxury markets were the focus of a wide range of garment types, including overcoats, suits, jackets, sweaters and various accessories such as scarves, hats and gloves. Overall, scarcity had created a luxury market for items containing camel fibres.

Llama fibres are from an animal native to the South American Andes, an area of high altitude, cool climate and low humidity. The animal has a fine undercoat, which offers protection from both excessive cold and heat. Fleece length and thickness vary greatly and are dependent on the climate and the frequency of shearing.

Vicuña fibres are from an animal of the same name and have been considered as the finest (and most expensive) animal fibres available. The animal is native to South America (particularly Argentina, Bolivia, Chile and Peru) and, by the late-twentieth century, was a protected species (under the International Convention of Trade in Endangered Species). It could, however, be trapped and sheared, unharmed, and then released, but could not be held in captivity. Vicuña fibres grow very slowly, are exceedingly soft, fine and warm, are cinnamon in colour and are rarely bleached or dyed. Fibres have been used in sweaters, shawls, coats and suits, often blended with relatively affordable cashmere fibres.

Yak fibre is from the animal of the same name. Found in Himalayan regions, the Tibetan plateau and some parts of East Asia, the animal has provided fibres for clothing, blankets, tents and ropes for several centuries. The fine, soft undercoat of fibres grown in winter offers substantial insulation properties. Like many other animal fibres, it outperforms sheep's wool in terms of warmth, softness and breathability. Overall, it is regarded as an exotic luxury fibre. Other exotic luxury fibres with seemingly similar properties to yak fibre are sourced from the bison (or buffalo) and musk ox (or qiviut). In both cases, an inner coat of fibres is offered, and this is soft, warm, light-weight, tough and washable.

Various further specialty hair fibres, such as beaver, stoat, squirrel and badger, were used in the past. It appears that often these were added to give perceived *caché* (and expense) rather than practical function.

Another animal fibre mentioned occasionally in relevant texts and websites is chiengora, fibre sourced from dog's hair and reputed to be much warmer and much less elastic than sheep's wool and with the ability to shed water well. Much of the commentary relating to the various exotic luxury fibres (especially chiengora) appears to be based on sentimentality rather than academic rigour; a need for thorough investigation is apparent.

Cultivated (or occasionally referred to as domesticated) silk is an animal fibre as well, though not a fibre type that is sourced from an animal's coat, but instead produced in the form of a spun cocoon of continuous fibres. Extruded from the front glands of a caterpillar (known as a silk worm), which feeds exclusively on the leaves of the mulberry tree, the cocoon (about half the size of a hen's egg) is spun by the insect as a protective environment around itself prior to emerging as a moth. However, if this metamorphosis is prevented by killing the insect within, the silk can be unravelled in continuous form. Silk fibres were extracted first, it seems, in ancient China and, by the early-twenty-first century, the People's Republic of China continued to hold a close association with the fibre though, by then, other locations were of importance also. Historically, in the European setting, the processing of silk was associated closely with Italy, France and Spain. In the British context, (like linen manufacture) stimulation was given to the expansion of silk processing during the late-seventeenth century when French Huguenot textile workers settled in Britain, initially in locations such as Spitalfields (London), as well as Southampton, Norwich, Canterbury and Bristol. In time, other British locations became noted centres for silk processing, among them Macclesfield, Coventry, Stockport and Congleton. By the end of the second decade of the twenty-first century, important international silk producers included the People's Republic of China, India, Brazil, Uzbekistan, Iran, the Democratic People's Republic of Korea (also known as North Korea), Thailand and Vietnam, with notable further levels of production coming from Bulgaria, Bangladesh, Japan and Turkey (the International Sericulture Commission website at https://inserco.org/en/).

Another important category of silk is the wild variety, historically grown in numerous countries, especially India (where it was known as vanya silk) and China (particularly Henan province). This silk type does not come in continuous-fibre form but, rather, as the insect is allowed to emerge from the cocoon in moth form and part of the cocoon is eaten away, the fibres are processed in discrete lengths (or staple form), and are generally referred to as wild silk (to differentiate it from cultivated silk).

When viewed longitudinally under the microscope, cultivated silk has a smooth surface and appears as a transparent rod. In terms of cross-section, the fibre is triangle shaped with rounded angles. Meanwhile, wild-silk fibres (which are in staple form) viewed longitudinally under a strong microscope appear broader with fine longitudinal lines on each and, in cross-sections, appear as flat wedges or spindle shapes.

Silk is very lustrous and very strong compared to most other natural fibres, but strength decreases on wetting. The fibre is extensible, but not as much as wool. Silk soils easily, absorbs water well, but dries slowly. Strong alkalis and acids weaken the fibres, as does chlorine bleach, and serious damage can result from the presence of perspiration. Further features are that the fibres are relatively expensive (compared to other fibres such as sheep's wool or cotton), and are non-flammable, but offer poor abrasion resistance. Relatively easy dyeing with a broad choice of dyestuffs is a positive feature.

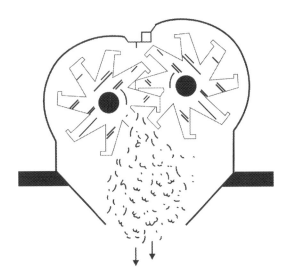

FIGURE 2.3   Scutching of flax. Developed from Hann (2005). Developed by JW.

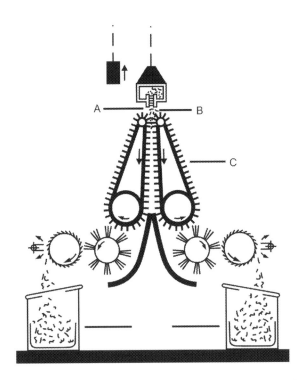

FIGURE 2.4   Hackling of flax. A = clamp; B = flax strands; C = comber pins (or needles); hackled tow waste in lower right-hand and left-hand boxes. Developed by JW.

Readers wishing to explore any of the avenues identified in this section should refer to the *Journal of Natural Fibres* (under this title since 2004) which has published numerous relevant articles. In the early-twenty-first century, the most comprehensive treatise concerned with natural fibres was provided in two volumes by Kozlowski (2012). Franck (2001) edited a useful treatise concerned with silk, mohair and other luxury fibres.

2.4   REGENERATED FIBRES

Fibres classed as regenerated are reconstituted from various natural substances (and this natural origin allowed their acceptance among traditional UK manufacturers in cotton-, wool- and flax-processing sectors, during the early- to mid-twentieth century). Although regenerated fibres have been formed from milk, peanuts, corn protein and seaweed, probably the most successful source has been various forms of wood pulp or similar cellulosic matter. The most important regenerated-fibre types have been viscose (developed in the early-twentieth century), acetate, triacetate, modal and lyocell. The greatest weakness of the early regenerated fibres was their lack

FIGURE 2.5   Silk cocoons. The University of Leeds.

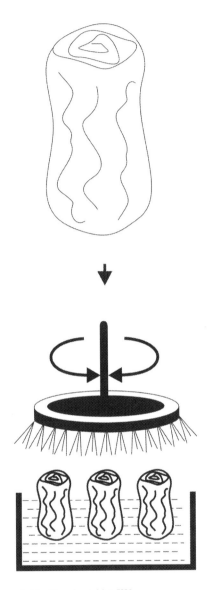

FIGURE 2.6 Extraction of silk. Developed by JW.

of strength particularly when wet. Relatively recent developments from the mid- to late-twentieth century include high-wet-modulus rayon (often referred to as modal), extracted from beech wood. The fibre is soft and shiny in nature with the ability to absorb more water than can be absorbed by cotton fibres and, although a variety of viscose, it is much stronger. Meanwhile, lyocell (developed in the late 1970s and 1980s) is similar in strength to polyester, stronger than cotton and in continuous-filament form can yield a silk-like woven textile. By the early-twenty-first century, lyocell was regarded as an environmentally friendly fibre, as the sources of cellulose

FIGURE 2.7   Early stage in the processing of staple silk (normally from several cocoons). Developed by JW.

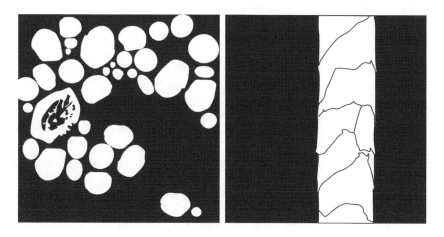

FIGURE 2.8   Microscopic view of merino wool. Drawn by JW.

FIGURE 2.9 Microscopic view of mohair. Drawn by JW.

FIGURE 2.10 Microscopic view of cashmere. Drawn by JW.

used were regarded as sustainable, and solvents used in manufacture were returned to the production system.

Viscose fibres are regenerated fibres consisting almost entirely of cellulose. Under a strong microscope, the fibre shows a serrated cross-section, and may be used in either continuous-filament or staple forms. Often viscose fibres are blended with other fibres. The fibres are highly lustrous and absorbent, although not very elastic. Viscose is not strong and loses substantial strength when wet. The fibre does not dry quickly and has a relatively smooth surface (which helps to resist easy soiling). Shrinkage may occur on washing, and the fibres are attacked by mildew, particularly if left damp. The fibres are not attacked by clothes' moths and are not damaged by weak alkalis (though strong alkalis cause swelling and weakening of fibres). Prolonged exposure to sunlight will also weaken fibres, as will excessive heat (so a

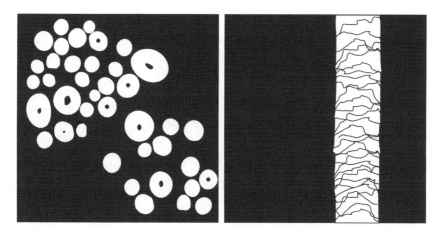

FIGURE 2.11   Microscopic view of camel. Drawn by JW.

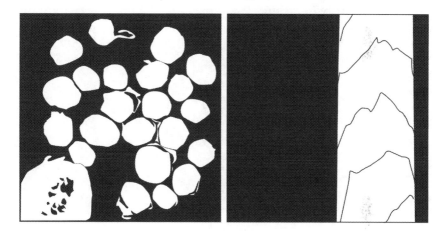

FIGURE 2.12   Microscopic view of vicuña. Drawn by JW.

cool iron is best). The fibres dye well with a variety of dyes and have proved popular in blends with other fibres.

Acetate is sometimes referred to as cellulose secondary acetate. The fibres are made from acetic acid and cellulose and resemble other cellulosic fibres to some degree. Under a strong microscope, the fibres show longitudinal ridges and a less serrated cross-section when compared to viscose. Acetate fibres are manufactured in continuous-filament form, although they are sometimes chopped, blended and processed in staple form, and are lustrous, soft and pliable. In cloth form, drape is good, and handle is regarded as attractive. The fibres are slightly stronger than viscose but, like viscose, lose strength when wet. Acetate is moderately elastic and has good crease resistance. The fibres do not shrink or soil easily but are not very absorbent. The fibres are thermoplastic, so require a cool iron after laundering. Resistance to mildew, insect and bacteria attack is high and the fibres are not discoloured by

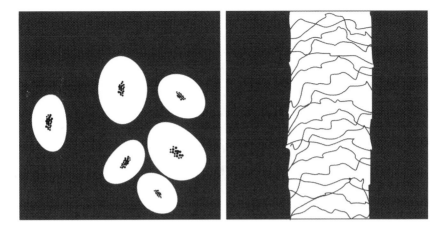

FIGURE 2.13 Microscopic view of llama. Drawn by JW.

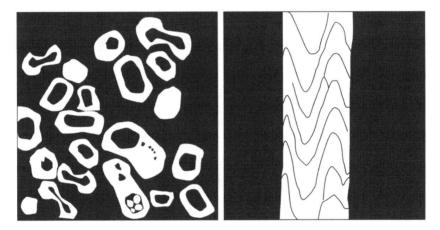

FIGURE 2.14 Microscopic view of rabbit. Drawn by JW.

sunlight; but prolonged exposure will encourage fibre weakening. The fibres are weakened also by strong acids and alkalis but are more resistant than viscose to dilute acids. Dyes suited to cotton and viscose are not used, but rather a class of dyes known as disperse dyes appear suited and are fast to washing and light. The fibres are cheap but more expensive than viscose and medium-grade cotton, and are also flammable, and not as durable as many synthetic fibres.

Triacetate is like acetate when viewed under a strong microscope, though the former is less absorbent than the latter. Triacetate dries very quickly after wetting and can be heat set (a thermal process, using steam or dry heat to impart dimensional stability). Compared to acetate, triacetate can be ironed at a slightly higher temperature and is slightly less flammable. Triacetate has a low absorbency, is difficult to dye (though, as with acetate fibres, disperse dyes are widely used), can be dry-cleaned, and deteriorates only slightly in strength when wet.

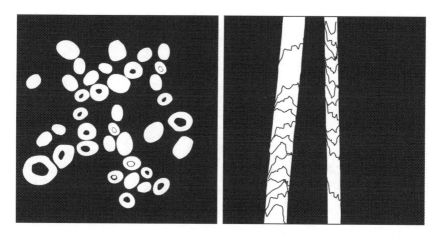

FIGURE 2.15   Microscopic view of beaver. Drawn by JW.

FIGURE 2.16   Microscopic view of silk (cultivated). Drawn by JW.

Modal, best regarded as a second-generation viscose, is a manufactured fibre consisting largely of cellulose. The fibres have high breaking strength, can absorb substantially more water than cotton and are stronger than polyester. High dimensional stability and low shrinkage are further features.

Lyocell is a generic term for a relatively recent generation of fibres derived from cellulose in wood-chip form (mainly, it seems, from closely managed plantations using land not suited to conventional farming). Though related to viscose, lyocell uses a solvent-spinning technique in which the cellulose does not undergo significant change. Lyocell requires a lesser number of production steps than viscose. The fibres have excellent wear resistance, are reputed to have very high bacterial resistance, are breathable, smooth, soft and comfortable, have high absorbency and good wicking ability (so keep the skin dry in hot weather), can be dyed to high standards, are very strong and are used in various industrial applications (in, for example,

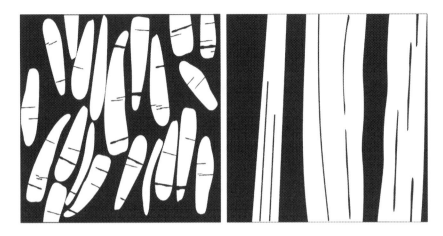

FIGURE 2.17   Microscopic view of silk (wild). Drawn by JW.

FIGURE 2.18   Microscopic view of cotton. Drawn by JW.

protective clothing or filters) as well as in a wide range of garment types. The major disadvantage appears to be the high cost compared to other fibres, and care must be taken in laundering. In the early-twenty-first century, lyocell fibres were regarded as eco-friendly and environmentally sustainable and were produced in a closed-loop process, which ensured that all solvents left after manufacture were recovered and returned to the process. In addition, it appears that the fibres are biodegradable. Woodings (2001) presented a comprehensive review of regenerated cellulosic fibres of various types.

2.5 SYNTHETIC FIBRES

Known also as chemical fibres, a name coined to generate market success, these fibres acquired a negative market image during the late-twentieth century due to their

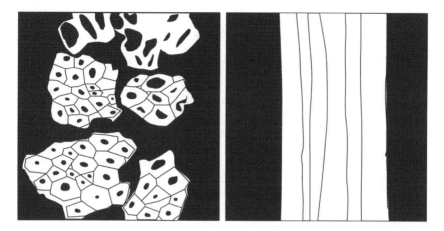

FIGURE 2.19 Microscopic view of hemp. Drawn by JW.

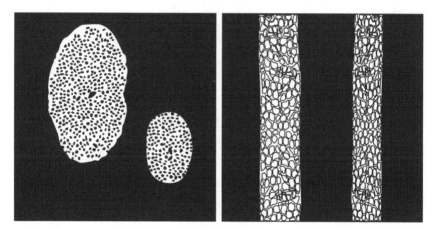

FIGURE 2.20 Microscopic view of coir. Drawn by JW.

association with the chemicals industry which had acquired a reputation as a major contributor to environmental pollution. There had also been a tendency (particularly during the late-twentieth century) to use synthetic fibres in inappropriate end uses; 100 per cent nylon sheeting and shirting are well-known examples and these helped to consolidate the negative views among consumers. Therefore, in general, by the late-twentieth century, synthetic fibres had acquired a negative image among the public. By the second decade of the twenty-first century, perceptions appeared to have changed and the processing of several forms of synthetic fibre was known to be less harmful to the environment than that of some natural fibres.

The discovery of nylon, the first synthetic fibre, occurred in the 1930s. Under strong magnification, nylon fibres appear smooth and even longitudinally, and with circular cross-sections (though these can be altered by changing the shape of the spinneret). Nylon fibres can be processed in continuous-filament or staple form. The

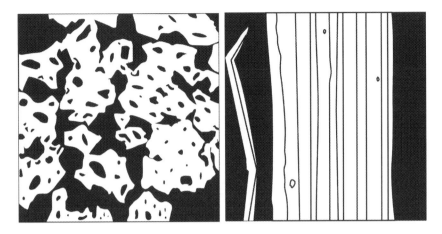

FIGURE 2.21   Microscopic view of jute. Drawn by JW.

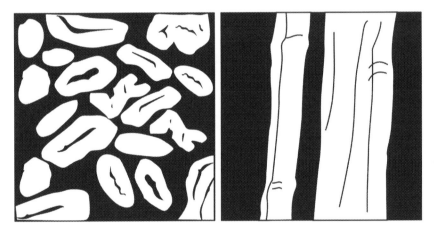

FIGURE 2.22   Microscopic view of ramie. Drawn by JW.

fibres have a bright lustre, but de-lustering treatments are possible. Nylon fibres are extremely strong with good abrasion resistance. Strength is not affected significantly by wetting (though the fibres do not absorb water well). Nylon fibres are very extensible and resistant to creasing. Insulation characteristics can be improved by using suitable fabric constructions. Fibres are resistant to soiling and alkalis, although they are weakened by acids. The fibres are flammable, easily stretched, weakened by prolonged exposure to sunlight, are not attacked by micro-organisms and are thermoplastic (so should be ironed at a very low setting).

In terms of quantities used, probably the most successful of all synthetic fibres has been polyester. Under strong magnification, polyester fibres are smooth and rod-like longitudinally. They may be used in continuous-filament or staple form. Fibres are highly lustrous but appear to be less glassy than nylon. Polyester fibres are very strong (though not as strong as nylon) and have good resistance to abrasion. Like

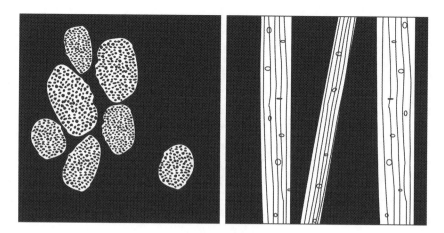

FIGURE 2.23   Microscopic view of abaca. Drawn by JW.

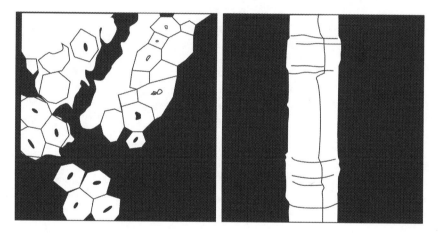

FIGURE 2.24   Microscopic view of flax. Drawn by JW.

nylon, strength does not deteriorate when wet. Polyester fibres are highly resistant to stretching, retain their shape well and do not crease easily. They have reasonably good soil resistance and are resistant to acids and alkalis. Polyester is thermoplastic and, when in cloth form, holds permanent creases well and resists deformation. The fibres are flammable, are not attacked by bacteria, mildew, or insects, and show some resistance to sunlight. Fibres are not easily dyed (but disperse dyes appear suited). Polyester has proved very popular in blended forms (with wool or cotton, for example). Deopura, Alagirusamy, Joshi and Gupta (2008) presented a wide-ranging review of polyester and polyamides, covering most aspects of processing and use.

Acrylic fibres appear smooth longitudinally under high microscopic magnification, and cross-sections vary with the method of manufacture, though dog-bone- or bean-shaped cross-sections are common. Used mostly in staple form, acrylic fibres are extensible, do not crease easily, are not as strong as nylon or polyester, but are

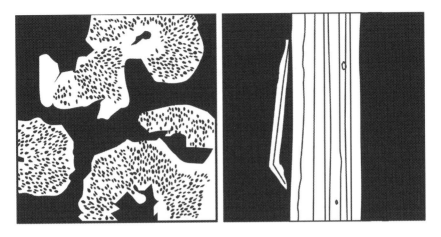

FIGURE 2.25   Microscopic view of sisal. Drawn by JW.

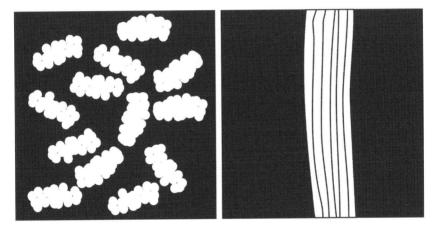

FIGURE 2.26   Microscopic view of viscose. Drawn by JW.

stronger than wool. Acrylic fibres are absorbent and regarded by many consumers as comfortable to wear, especially when knitted. Garments from acrylic fibres do not soil or stain easily, and are highly resistant to acids, alkalis, bleaches and solvents used in domestic or commercial laundries. Fibres are thermoplastic and can be heat set (as noted briefly in section 2.4, this is a thermal process, which uses steam or dry heat to impart dimensional stability), and are not attacked by bacteria, mildew or insects. Acrylic fibres are resistant to sunlight, are very flammable (producing poisonous hydrogen cyanide when burning) and quite cheap when compared to low-quality wool. Modacrylics are less flammable, are often used in blends as furnishings (including curtaining and upholstery) and, because of their soft handle, are used often in synthetic fur.

Olefins (sometimes referred to as polyolefins) are a category of synthetic fibre patented in the 1950s. Probably the most renowned of olefins are polypropylene and

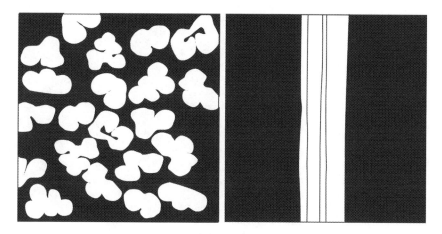

FIGURE 2.27   Microscopic view of acetate. Drawn by JW.

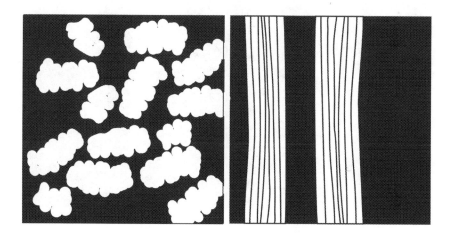

FIGURE 2.28   Microscopic view of triacetate. Drawn by JW.

polyethylene, both used as textile fibres, with the major advantage of strength (wet or dry), as well as resistance to sunlight, mildew, staining and abrasion. With a low melting point, the fibres can be thermally bonded with ease. Ugbolue (2009) provided a well-focused review.

Elastane fibres, first synthesised in the early 1950s in Germany by Bayer Fibres, are exceedingly extensible. Occasionally, during the early-twenty-first century, elastane was used on its own, but, more commonly, it was combined with other fibres. Elastane with nylon, used often in swimwear, was a common combination. Elastane fibres are readily dyeable with a wide range of dye possibilities and have a good resistance to oils and suntan creams. The fibres have good abrasion resistance and have been used frequently in foundation garments, swimwear, support hosiery, sports and other leisure wear.

FIGURE 2.29   Microscopic view of modal. Drawn by JW.

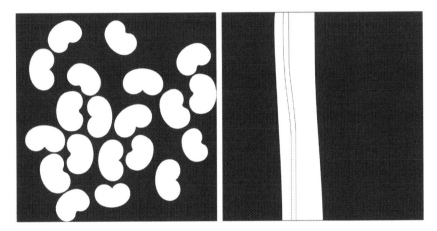

FIGURE 2.30   Microscopic view of acrylic. Drawn by JW.

Glass fibres, often used for domestic insulation, are exceedingly strong, though with high rigidity they are also very brittle. Due to their high strength and relatively low weight, the fibres have been used extensively as reinforcing additions to composites of various kinds in numerous industrial sectors.

Aramid fibres are synthetic fibres, characterised by the potential for high performance with respect to low weight, strength and resistance to heat. Known commonly by the DuPont brand name, *Kevlar*, categories of these fibres were used commonly as reinforcing components in composites, as well as for applications in aerospace, ballistics, military and sports areas. They have functioned as the main component in protective clothing where they have offered protection against heat and chemicals.

Carbon fibres, a further class of fibres which are exceedingly strong, are used in various high-performance areas; the strongest varieties of the fibre have a tensile strength five times that of steel. As well as high strength, further advantages include

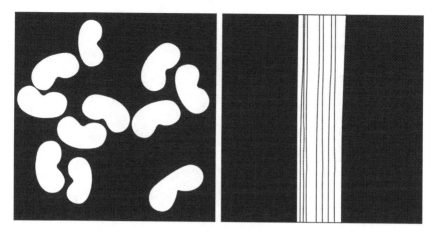

FIGURE 2.31 Microscopic view of modacrylic. Drawn by JW.

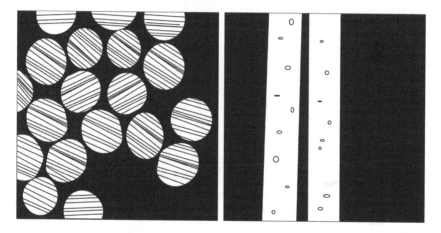

FIGURE 2.32 Microscopic view of aramid (Kevlar). Drawn by JW.

high stiffness, relatively low weight, high tolerance to a broad range of chemicals and high-temperature resistance. Metallic fibres are fibres from any metal. In the early-twenty-first century, generally the fibres were flat in nature and covered with a clear plastic film. Another possibility was to have a core (of say silk) and to twist a metallic tape around this core. It seems that metallic fibres were used sparingly for embellishment purposes only.

As mentioned in the introduction to this chapter, the most renowned twentieth-century publication concerned with manufactured fibres was volume two of the two-volume treatise produced by Cook (1993b, [1984b]). This stands as the most comprehensive survey of manufactured fibres available at the time but has been supported well by publications since, including Klein (1994), Houck (2009) and both volume 1 and volume 2 of the *Handbook of Textile Fibre Structure* edited by Eichhorn, Hearle, Jaffe and Kitutani (2009). Late-twentieth-century fibre types were

FIGURE 2.33 Cotton bolls. The University of Leeds.

reviewed well by Hongu and Phillips (1997). Later, in 2005, Hongu, Phillips and Takigami presented an excellent review of the range available.

2.6 SUMMARY

This chapter has identified the main natural- and manufactured-fibre types. Natural fibre types include cotton, extracted from seeds, flax, jute, hemp, ramie, kenaf, nettle, bamboo and banana, all taken from within the stems of plants or trees and known as bast fibres, and sisal, abaca and pina, the main leaf-fibre types. Brief mention was made of various fibres of mineral origin, including asbestos, glass, metallic, carbon and ceramic fibres. The main animal fibres were identified also, including wool from domesticated sheep and various 'specialty hair fibres' such as cashmere, mohair, camel and alpaca, as well as both domesticated and wild silk. Among the manufactured fibres, identification was made of regenerated fibres such as viscose, acetate, triacetate, modal and lyocell, and synthetic fibres such as nylon, polyester, acrylic, modacrylic, polypropylene, elastane, aramid and carbon.

REFERENCES

Australian Government (2008), *The Biology of Gossypium Hirsutum L. and Gossypium Barbadense L. (cotton)*, Melbourne: Australian Government Office of the Gene Technology Regulator.

Barber, E. J. W. (1991), *Prehistoric Textiles*, Princeton, NJ: Princeton University Press.

Beckert, S. (2014), *Empire of Cotton. A Global History*, New York: Knopf.

Cook, J. G. (1993a, [1984a]), *Handbook of Textile Fibres (I, Natural Fibres)*, Durham, UK: Merrow.

Cook, J. G. (1993b, [1984b]), *Handbook of Textile Fibres (II, Man-made Fibres)*, Durham, UK: Merrow.

Deopura, B. L., R. Alagirusamy, M. Joshi and B. Gupta (eds.) (2008), *Polyesters and Polyamides*, Cambridge, UK: Woodhead.

Dirgar, E., and O. Oral (2014), 'Yarn and Fabric Production from Angora Rabbit Fiber and Its End-Uses'. *American Journal of Materials Engineering and Technology*, 2(2):26–28.

Eichhorn, S., J. W. S. Hearle, M. Jaffe and T. Kikutani (eds.) (2009), *Handbook of Textile Fibre Structure, Volume 1: Fundamentals and Manufactured Polymer Fibres, and Volume 2: Natural, Regenerated, Inorganic and Specialist Fibres*, Cambridge, UK: Woodhead.

Franck, R. R. (ed.) (2001), *Silk, Mohair, Cashmere and Other Luxury Fibres*, Cambridge, UK: Woodhead.

Franck, R. R. (ed.) (2005), *Bast and Other Plant Fibres*, Cambridge, UK: Woodhead.

Gordon, S., and Y. L. Hsieh (eds.) (2006), *Cotton. Science and Technology*, Cambridge, UK: Woodhead.

Hann, M. A. (2005), *Innovation in Linen Manufacture*, a monograph in the Textile Progress series, 37(3):1–42.

Hearle, J. W. S., L. Hollick and D. K. Wilson (2001), *Yarn Texturing Technology*, Cambridge, UK: Woodhead.

Hongu, T., and G. O. Phillips (1997), *New Fibres*, second edition, Cambridge, UK: Woodhead.

Hongu, T., G. O. Phillips and M. Takigami (2005), *New Millennium Fibres*, Cambridge, UK: Woodhead.

Houck, M. M. (ed.) (2009), *Identification of Textile Fibres*, Cambridge, UK: Woodhead.

ICTSD (2013), *Cotton. Trends in Global Production, Trade and Policy; Information Note*, Geneva, Switzerland: International Centre for Trade and Sustainable Development.

Johnson, N. A. G., and I. Russell (eds.) (2008), *Advances in Wool Technology*, Cambridge, UK: Woodhead.

Klein, W. (1994), *Man-made Fibres and Their Processing*, Cambridge, UK: Woodhead.

Kozlowski, R. (ed.) (2012), *Handbook of Natural Fibres, vol. 1: Types, Properties and Factors Affecting Breeding and Cultivation and vol. 2: Processing and Applications*, Cambridge, UK: Woodhead.

Sharma, H. S. S., and C. F. Van Sumere (1992), *The Biology and Processing of Flax*, Belfast, N. Ireland: M publications.

Simpson, W. S., and G. Crawshaw (eds.) (2002), *Wool. Science and Technology*, Cambridge, UK: Woodhead.

Ugbolue, S. C. O. (ed.) (2009), *Polyolefin Fibres*, Cambridge, UK: Woodhead.

Woodings, C. (ed.) (2001), *Regenerated Cellulosic Fibres*, Cambridge, UK: Woodhead.

WEBSITE REFERENCES

Food and Agricultural Organisation of the United Nations, website at: http://www.fao.org/economic/futurefibres/home/en/ (accessed at 9-30am, 10 February 2020).

International Sericulture Commission, website at: https://inserco.org/en/ (accessed at 3.30pm, 9 September 2019).

3 Yarns

3.1 INTRODUCTION

The purpose of this chapter is to present an outline of methods used in the formation of continuous twisted structures known as yarns. Although it is accepted that some yarns may be produced in tape form (and do not pass through what could be referred to as a traditional processing sequence), these are largely ignored within this book as they constitute only a very small proportion of the total yarn output worldwide.

All manufactured fibres are continuous in nature but after extrusion require stretching and twisting to impart the strength needed for further processing, cloth formation and use. After stretching, if desired, such fibres can be cut into a range of lengths (known as staples) and can be processed further on standard yarn formation machinery (used, for example, in cotton-, flax- or wool-yarn manufacture) (Figure 3.1). Other than cultivated silk, all common natural fibres come in staple-length form and require further processing in several stages. The initial intentions are to clean, to blend and to create a non-twisted continuous rope form known as a sliver (with numerous overlapping fibres in cross-section). As this sliver undergoes further processing, impurities and extra-short fibres are removed, remaining constituent fibres are made as parallel as possible, regularity of the sliver along its length is imposed and, after further attenuation, the structure is twisted continuously along its length with the aim of creating a yarn. As noted above, several stages of processing are required to achieve this final aim, and brief explanations are given below of the more important of these. It should be noted that, although different processing sequences have evolved to cope with different fibre types, the aims of removal of impurities and extra-short fibres, sliver formation, parallelisation of fibres, regularity (similar density of fibres throughout), attenuation of slivers and final twisting of fibres remain the same across processing systems associated with different fibre types. As a rule, across processing systems, it is the case that the degree of twist given to fibres will determine the nature of the yarn product; loose twist produces a softer yarn, and tight twist, a hard-twisted structure (known as a crêpe yarn). Twist may be either clockwise (i.e. in S direction, when the yarn is held vertically and viewed along its length, the direction of the twist follows the central bar of the letter 'S') or anti-clockwise (i.e. in Z direction, when the yarn is held vertically and viewed along its length) (Figure 3.2).

The use of the terms 'woollen' and 'worsted' appears to have widespread applicability; often, however, neither is necessarily associated directly with wool manufacture. Rather, each term is used to refer to a type of spinning sequence or system, with the woollen system less rigorous and using shorter fibres, and the worsted system more extended and using longer fibres; the former is associated with a looser, bulkier yarn and the latter with a tightly twisted, highly regular and smooth yarn. The term 'semi-worsted' indicates a hybrid system, which incorporates aspects of

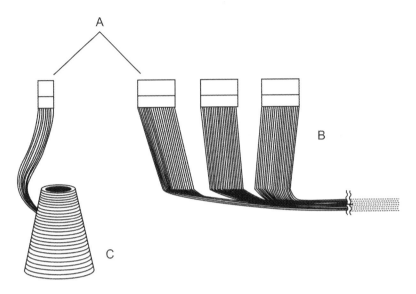

FIGURE 3.1 Manufactured fibres (A) may be cut and processed as staple yarn (B) or may be spun as continuous-filament yarn (C). Re-drawn by JW.

both woollen and worsted systems. A further system, known as the 'cotton system', is used worldwide to process most staple yarns. A figure of 90 per cent of the total staple-spun yarns using the cotton system was cited by Tausif, Cassidy and Butcher (2018: 36). Variations and adaptations of the three systems (woollen, worsted and cotton), with the occasional addition of highly specialist equipment, are common; the principal characteristic to consider is the average staple length of the fibres being processed and the intended outcome. In short-staple spinning, based largely on the cotton system, different routes are possible, producing combed yarn or carded yarn, both using a system known as ring spinning, or yarn spun by an alternative system, such as rotor spinning (which requires fewer processing stages). Alternatively, staple spinning (often of longer fibres) can go down either the woollen-yarn system or the rather more extended worsted system, or, the fibres in the form of continuous-filament man-made fibres, the fibres can be stretched and broken and blended within the worsted system. A useful flow chart of the various routes was provided by Tausif, Cassidy and Butcher (2018: 36).

All systems of yarn manufacture are governed by what is known as count, which denotes the physical size or linear density of a yarn. Yarn counts are based either on weight per unit length, known as direct counts, or on length per unit weight, known as indirect counts. Most traditional spinning systems developed their own count systems, specific to the yarns produced. There are, for example, count systems which were developed in close association with traditional spinning systems in the British Isles to denote the linear density of woollen, worsted, cotton, linen or silk yarns. Probably the most common yarn-count systems used worldwide are denier and tex, both direct-count systems. Denier is the weight in grams of 9,000 metres of yarn (or

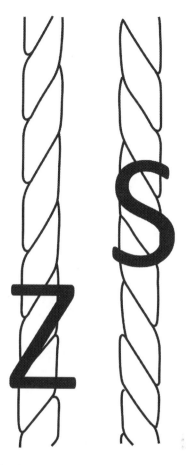

FIGURE 3.2   Yarn with Z twist and yarn with S twist. Drawn by JW.

filament), and tex is the weight in grams of 1,000 metres (or 1 kilometre) of yarn (or filament).

A typical outline of processing (with initial stages shown in Figure 3.3) for cotton fibres arriving in bale form at the mill could include: a blow-room process, which further cleans and opens the cotton fibres, removing dust, seed particles and other non-fibrous material. Next, the fibres undergo a process known as carding, which removes extra-short and entangled fibres and delivers either a wad of fibres known as a lap (suited for nonwoven cloth manufacture) or a continuous untwisted rope form known as a sliver, in which constituent fibres exhibit a small degree of parallelisation with each other; this sliver, together with several others, is passed (at least once) through a machine known as a drawing frame (Figures 3.4, 3.5 and 3.6) which helps to make the sliver more regular (with approximately equal numbers of fibres in cross-section throughout) and also attenuated (or drafted); a further stage of attenuation, using a machine known as a speed frame produces a thin sliver known as a roving (Figure 3.7) which is then passed to a spinning frame (Figures 3.8 and 3.9)

FIGURE 3.3 Cotton fibres from bales, blended and carded to make sliver (in sliver can to right). Various alternatives are possible, depending on pre-mill processing. For example, a blow-room process may be required prior to blending of fibres. Re-drawn by JW.

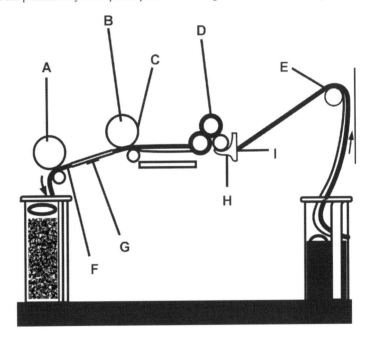

FIGURE 3.4 Doubling and drafting (or drawing) frame. A = delivery rollers; B = drafting rollers; C = drafting guides; D = retaining rollers; E = guide roller; F = delivery guides; G = doubling plate; H = feed guides; I = back feed guides. Developed by JW.

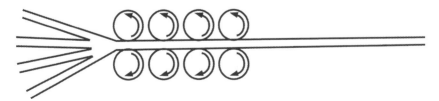

FIGURE 3.5 Drafting and doubling (four slivers to one). Re-drawn by JW.

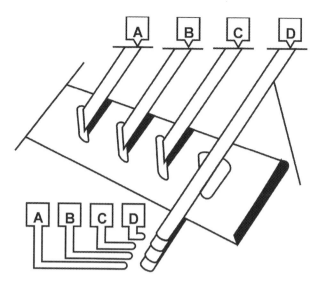

FIGURE 3.6 Doubling plate A, B, C, D (four slivers to one). Developed by JW.

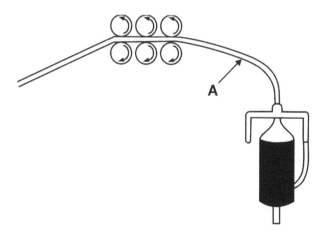

FIGURE 3.7 Sliver drafted to roving (A). Re-drawn by JW.

where, after further attenuation, the structure is twisted to provide a yarn. Yarn production can be enhanced through the addition of further machine processing prior to
yarn formation; this possible additional processing may be aimed at the further parallelisation of fibres and the improved regularity of the sliver. In addition, a choice
of spinning systems may be available, though by the early-twenty-first century most
industrially spun yarns globally were ring spun. Further details of the different processing stages are provided in the sections below. It is assumed here that all staple-
fibre processing begins with carding, though with different types of fibre there may
be further preliminary (or preparatory) stages.

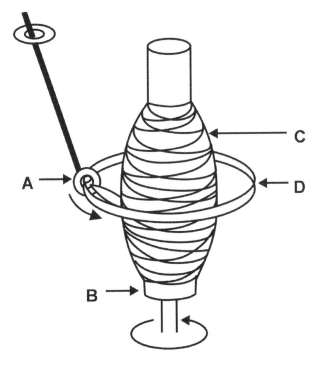

FIGURE 3.8   Ring spinning (A = traveller; B = bobbin on spindle; C = bobbin with yarn; D = ring). Developed by JW.

3.2   FILAMENT, STAPLE AND OTHER YARN TYPES

As noted in the introduction to this chapter, other than cultivated silk all common natural fibres come in staple-length form and require further stages of processing before conversion to a continuous twisted state. A staple-spun yarn is a largely regular lengthwise assembly of broadly parallel and overlapping fibres, held in association often through the inserted twist, and thus forming a continuous rope-like form, with close to equal numbers of fibres in cross-sections along its length. Processing can be intensive depending on the anticipated end use of the yarn. Various categories of staple-spun yarn can be identified often by reference to the intensity of processing and the type of spinning technique used (Thompson and Thompson, 2014: 56).

As noted previously (in the introduction to this chapter), the essential aims of staple-fibre processing, no matter which system, are similar, though differences have evolved across all traditional systems over the years. Stages in preparation for spin-ning vary across systems, as often the exact impurities to be removed are different. There are numerous potential ways of classifying or grouping yarns, and the system adopted below has been developed from the Textile Institute's publication *Textile Terms and Definitions* (Tubbs and Daniels, 1991). Within this, yarns were classified simply as staple, singles, folded (double or plied), cord, continuous-filament, stretch,

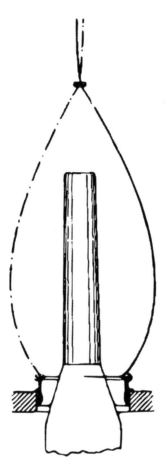

FIGURE 3.9 Balloon shape during ring spinning. Developed by JZ.

texturised (sometimes spelled textured), high-bulk, bi-component, fancy (as well as metallic) or specialist yarns.

Staple yarns are yarns manufactured from a 'mass of fibres having a certain homogeneity of properties, usually length' (Tubbs and Daniels, 1991: 296). Yarns may be singles, folded or corded. The term 'singles' denotes a single component yarn. The term 'folded' (or occasionally 'doubled') may be used where two (or more) yarns are combined through twisting, and the term 'corded' to denote the combination of two or more folded yarns. Continuous-filament yarn can be described as a yarn 'composed of one or more filaments that run essentially the whole length' (Tubbs and Daniels, 1991: 67). Stretch yarns are characterised by a capacity to stretch and to recover fully from stretch (Tubbs and Daniels, 1991: 301). When used in the context of cloths, the term 'stretch' refers to 'materials with greater extension and recovery properties than traditional woven or knitted structures' (Tubbs and Daniels, 1991: 300). Texturised (mentioned briefly in section 2.5) is a term applied largely to thermoplastic, continuous-filament yarn that 'has been processed to introduce durable crimps, coils, loops or other fine distortions along the length of the filaments' (Tubbs and Daniels, 1991: 314). A texturised yarn is therefore a filament yarn endowed with crimped, looped or coiled effects, and is often bulkier than the equivalent non-texturised yarn. The term 'high-bulk' relates to the linear density or volume of a yarn. A bulkier yarn has greater covering power than a less bulky yarn. A high-bulk effect can be obtained by blending fibres of different shrinkage potentials and allowing shrinkage after spinning. A bi-component yarn consists of two or more components, invariably with the different components offering unique properties. Fancy yarns (or novelty yarns) may be characterised by periodic additions of one kind or another and may be further enhanced through twist variation or by changes in the feeding rate of one of the component yarns. The use of metallic-type fibre is often a feature of fancy yarns. Specialty yarns may also be created with specific performance characteristics in mind. Examples include sewing threads or yarns used in tyre cords. In years past, when natural fibres dominated, rope manufacture was of importance to many economies worldwide. The technology used was often influenced by innovations in other types of fibre processing. McKenna, Hearle and O'Hear (2004) provided a comprehensive handbook of the technology associated with rope manufacture; much of this is of relevance to yarn manufacture in general.

It is at the stage of opening (a typical preliminary action associated with a traditional cotton sequence) that available batches are taken and blended, by selecting quantities of fibre from different sources. As clumps are separated, air-operated cleaning machines help to remove non-fibrous matter. After this opening and cleaning, the fibres are compressed uniformly into a batt or thick sheet (with approximately constant thickness throughout), in preparation for carding. Carding helps to clean the fibres further, remove extra-short fibres and combine the remaining fibres into a uniform web, which can then be condensed into a continuous untwisted rope form or sliver. This continuous rope form is placed in a sliver can and presented to a drawing stage. Drawing encourages parallelisation of the constituent fibres. Often, several slivers are passed through the machine simultaneously, thus introducing a degree of further blending and, through a speed differential (involving faster rollers at the output end of the machine compared to those at the input end), the slivers are attenuated or drafted. Invariably, slivers are passed through two drawing frames to ensure a relatively uniform mass of straight and parallel fibres with broadly similar mass per unit length throughout the sliver. Two passes, of course, lead to greater expense. Rather than a second drawing stage, the fibres may instead be presented to combing, with further subsequent drawing and roving, prior to spinning and winding. An especially fine and regular yarn is achievable through the inclusion of combing, in which the output sliver from drawing is presented to the comb and pulled in sections through a series of finely spaced teeth, removing further non-fibrous material as well as entangled and extra-short fibres. After overlapping of resulting sections, a cleaner, more uniform sliver with parallel constituent fibres results.

From drawing and/or combing the cotton slivers pass to the speed (or roving) frame, which attenuates the slivers further into a form known as a roving, in preparation for ring spinning; as mentioned in section 3.1, by the end of the second decade of the twenty-first century, ring spinning was the most common spinning technique worldwide. Ring spinning reduces the roving mass even further and winds the twisted yarn on to a bobbin which then must be rewound (through a process known as winding) on to a package (yarn container or holder) suited for further processing. However, by the early-twenty-first century, ring spinning was regarded as relatively slow, particularly when compared to rotor spinning which could take the sliver from drawing and produce yarns wound on to a full-sized package (of the type associated with winding). Rotor spinning was thus substantially faster (up to seven times the speed of ring spinning) and could eliminate both the roving and the winding stages; the major disadvantage was that yarns were weaker, and the range of yarn counts achievable by rotor spinning was severely restricted. Another system, known as air-jet spinning, could operate also from the sliver rather than roving stage. By the end of the second decade of the twenty-first century, operational speeds of air-jet spinning were up to twenty times those for ring spinning, but yarns were weak and within a narrow count range. A further spinning possibility by the end of the second decade of the twenty-first century was friction spinning, but substantial further research and development appeared to be required to ensure dependable operational speeds and adequate yarn quality across a wide product range.

3.3   BLENDING, CLEANING AND SLIVER FORMATION

An important early stage in the processing of fibres is known as carding. As noted briefly in section 3.2, carding helps to clean an assembly of fibres and removes extra-short fibres. The process is achieved using a carding engine or card. Different types of card are associated with different spinning systems. Invariably, across all spinning systems, the objectives of the process are to disentangle fibres, to remove non-fibrous matter, to remove extra-short fibres and to form an arrangement known as a lap (a sheet of fibres, of close to regular thickness) or a sliver (a continuous assembly of overlapping fibres). Typically, a card consists of a rotating cylinder covered with bent pins (known as card clothing), surrounded by a series of smaller cylinders also rotating and covered with bent pins. When a wad of fibre is passed between the larger cylinder and one of the smaller cylinders, clumps of fibre are broken up, and non-fibrous matter, together with extra-short fibres, fall away. At the exit point of the machine, the wad (by then much cleaner) may be presented in that form or else may be presented in a continuous sliver form. While due to the directional movement and orientation of the card a small degree of parallelisation of fibres within the sliver may occur, substantial further parallelisation may be required prior to spinning of the fibres into yarn form.

Another important function of carding is that it gives an entry point for blending of different fibres; although blending may be achieved at various stages during processing, entry at the carding stage is probably the most common. Blending is a combination of two or more fibre types, invariably with the intention of lowering costs or creating a combination with attributes greater than any one of the constituents if taken on their own. No fibre is perfect for all uses. Some are good and some poor. Blending offers the opportunity to combine qualities to achieve a more desirable result. The aim is to combine fibres of different origins, and to ensure that each fibre type is represented to an equal degree in any cross-section of the final yarn. For example, characteristics such as strength or elasticity, or the combination of expensive with less expensive fibre types, may be desirable. It is wise, of course, even when the intention is to produce yarns from only one fibre type, to use collections of fibres from different sources (so that any weaknesses will be averaged out). Cotton, for example, may arrive in the mill/factory in bale form. Bales will need to be opened, fibres taken and weighed. At the blow-room and/or carding stages, blending of fibres from different bales/batches would be the norm.

As noted previously (in section 3.2), when exceedingly regular and clean yarns are the intention, a process known as combing can be used. On entering the combing machine, a section of the fibres is pulled out and passed through a series of metal teeth which help to ensure that the fibres within are parallel to each other. This and other sections of cleaned fibres are overlapped with each other in a long line (in reconstituted form). At the same time, dust and other non-fibrous matter as well as extra-short fibres are removed. Across textile industries, the short fibres from combing waste are regarded often as high-quality waste; in the worsted industry, for example, they are known as noils, and are invariably recycled to produce carded yarns. Meanwhile the combed sliver can then be used to spin highly regular yarns. Carded or combed slivers are then passed to the subsequent stages of processing.

3.4   DRAWING, DOUBLING, ROVING AND SPINNING

Drawing (or drafting) is the attenuation of a sliver, achieved through rollers moving more quickly at the front (the output end) of a machine compared to the back (the input end). Following the idea that thick and thin patches will often coincide when different slivers are simply placed one on top of the other, doubling is aimed at achieving a more regular sliver. As noted previously (in section 3.2), the term 'doubling' can be used also to refer to the action associated with combining two or more yarns into one twisted structure. In the context of slivers, doubling is often combined with drafting. So, if four slivers (each roughly equal in terms of weight per unit length) are fed into a machine (known as a draw frame) with input rollers turning at a speed equal to, say, x metres per minute and the output rollers (or similar arrangements) are turning at, say, $4x$ metres per minute (thus, considered to have a draft of 4) and within the machine the four slivers are brought together one on top of the other, then the resultant single output sliver will be broadly the same weight per unit length as each of the input slivers but will be more regular. Therefore, in this case, doubling and drawing have served the purpose of regularity and attenuation, respectively.

An important addition to yarn-preparation machinery, designed to enhance regularity, is the autoleveller; the attachment is generally added at the drawing stage but can be attached to a carding machine also. With autolevelling, the linear density (or weight per unit length) is measured and the draft (or degree of attenuation) adjusted accordingly, during processing. The aim is to address linear density variation by adjusting the draft before the final sliver is delivered from the machine. When the linear density is judged to be too high, the entry speed can be reduced, or if the linear density is judged to be too low, the entry speed is increased. With either approach, the exit velocity remains the same, as does the speed of production. Advantages include the production of a more regular yarn with fewer yarn breaks during cloth formation, which improves cloth quality and leads to both increased labour and machine productivity.

In all spinning systems, doubling and drawing will achieve higher regularity of slivers, with further attenuation through the speed (or roving) frame. These rovings (with slight twist) are wound on to bobbins suited to further processing into yarns. Various spinning frames have evolved over the years, initially suited to a restricted fibre type or average staple length and best used within a range of yarn counts; typically, however, through further minor innovations, the fibre applicability of relevant machines has been extended and the suitable count range broadened.

By the beginning of the third decade of the twenty-first century, the two most common spinning techniques used worldwide, within industrial settings, were ring spinning and rotor spinning, with the former substantially more common than the latter. Ring spinning necessitated roving and winding stages, whereas rotor spinning allowed for the omission of these. While ring spinning allowed a wide range of counts, from relatively fine to relatively coarse, rotor spinning was still restricted to the production of relatively coarse products only. It seems likely that the widening of the count applicability of rotor techniques will be a primary economic driving force in the first half of the twenty-first century. The substantially higher production speeds possible with rotor-spinning techniques, together with the possibility of leaving out roving as well as winding from the production sequences, will be irresistible factors to further innovation aimed at extending the applicable count range of rotor techniques.

Other important developments should be recognised. These include air-jet spinning and friction spinning, both with exceedingly high delivery rates but in each case producing yarns that are regarded as bulky, of low strength (a little lower than rotor-spun yarns and significantly lower than ring-spun yarns) and with non-uniform twist along the length of yarns. End uses range from blankets to rags and mops for cleaning, as well as filtration cloths, some leisure wear, upholstery and tablecloths. Other spinning systems worthy of mention include false twisting and stufferbox but, by the early-twenty-first century, each required extensive technological development to gain broader product applicability.

3.5   TWISTING, WINDING AND FANCY YARNS

The cop or bobbin which contains spun yarn from ring spinning (the most-used system) is not suited for further processing, so the yarn needs to be rewound on to a more suitable package (or yarn holder) through a process known as winding. The intention is to create a larger yarn package (possibly known as a cone or cheese) from several of the smaller cops. The yarn can also be checked for spinning defects, and these may be removed. If required, the yarns may be waxed during the process.

In precision winding, successive coils of yarn are placed closely adjacent and parallel to each other. A dense package, with the maximum length of yarn within the space taken, results. However, the package is not stable and supporting flanges are required to ensure that the packages of stacked yarn do not deteriorate. A further disadvantage of precision-wound packages is that the rate of unwinding is low. Non-precision winding (or drum winding), on the other hand, provides packages which, although they hold less yarn compared to precision-wound packages, are inherently stable, do not require supportive flanges and have a high rate of unwinding. Non-precision winding is the common choice for staple yarns across the textile industry worldwide. As noted above in the first paragraph of this section, the removal of yarn flaws is possible at the winding stage. Devices to remove such flaws may be either simple mechanical slits or digital in nature. The former will detect and prevent thick parts (often called slubs) from being wound; the machine will stop, and an operative will remove the slub and tie a knot to connect the two ends of the remaining yarn. A further aspect of mechanical devices, because yarns need to pass physically through the opening between two pieces of metal, is that they abrade the yarn to some degree. Digital clearers, on the other hand, will not abrade the yarn and will detect not only slubs but also thin yarn areas.

Twist is of great importance. It binds the constituent fibres together and, if optimised, will impart the highest possible strength to the yarn. Twist can be measured simply by the number of twists per unit length, and, as explained previously (in section 3.1), the twist direction can be described as 'S' or 'Z', depending on the direction of fibres when the yarn is held vertically. Once spun, yarns of the same type are often combined through twisting; as indicated in section 3.2, the term 'folding', or occasionally 'doubling', is used to refer to the process. Folding of yarns leads to improved uniformity, greater abrasion resistance and higher tenacity. The twist of a folded yarn should be less than that for each of the constituent yarns and the direction of twist should be the opposite of that of the constituent yarns. So, as a rule, if the constituent yarns are 'S' twisted, the folded twist should be 'Z'.

A category of yarns is worth further explanation. This group, known as 'fancy' (or, occasionally, as 'novelty' or 'effect') yarns, with each consisting of two or more components (often with one or more yarns circling around another), was mentioned briefly in section 3.2. Fancy yarns may be used as effect yarns in weaving or knitting and possibly also in embroidery and crochet. They are used often in small quantities to impart embellishment to high-value items, maybe to textiles in lounges, lobbies and elsewhere in the corporate sector, or possibly in some exclusive garment, where they are deemed to add *caché*. So, fancy yarns are considered to add a further aesthetic dimension to a textile.

Early forms of fancy yarn may well have evolved from yarn imperfections of one kind or another being perceived as a good source for occasional embellishment of woven or knitted textiles. Subsequently, attention may have been focused on how best to plan these imperfections, with enquiry initially based on combining one effect with another. Gong and Wright (2002: 2) defined fancy yarns as 'those in which some deliberate decorative discontinuity or interruption is introduced'. Fancy yarns therefore feature interruptions or discontinuities along their length; these may be associated with changes in colour and/or texture. By the early-twenty-first century, relatively complex machinery was in use, combining components, maybe consisting of one or more core parts, with further yarns, sliver or roving wrapped around. A good review of fancy-yarn manufacturing techniques was provided by Gong and Wright (2002).

Fancy yarns are classified in various ways, with the following types characteristic in any classification. A marl yarn is regarded as the simplest fancy yarn, made by simply doubling two yarns of similar count and twist but of different colours and/ or textures. A spiral or corkscrew yarn is, typically, where one yarn spirals around another; often yarns are of different twist and thickness. Various examples were described by Gong and Wright (2002: 34–35). A gimp yarn has a twisted core yarn with an effect yarn wrapped around it, producing wavy projections on its surface (Tubbs and Daniels, 1991: 110). Gong and Wright (2002: 40–41) explained that loop yarns consisted of projections which were almost circular in nature and involved the combination of four yarns. Snarl yarns display 'snarls or kinks projecting from the core' (Tubbs and Daniels, 1991: 112). They are manufactured by the same procedure as a loop yarn (described above), but, as stated by Tubbs and Daniels, 'instead of a resilient [core] … a lively highly twisted yarn is used' (Tubbs and Daniels, 1991: 112). Snarls are therefore formed in place of loops when the tension of the front rollers is released. Snarls may be controlled in order to vary their size and frequency, 'either continuously or in groups at places along the yarn' (Tubbs and Daniels, 1991: 112). Knop yarn (also known as nepp yarn) is yarn made on the woollen-spinning system. The yarn has 'strongly contrasting spots on its surface that are made either by dropping in small balls of wool at the latter part of the carding process or by incorporating them in the blend and so setting the carding machine that these lumps are not carded out' (Tubbs and Daniels, 1991: 110). Further description was provided by Gong and Wright (2002: 45–46). Slub yarns are those yarns where slubs (or thickened parts) have been created deliberately to give an effect of discontinuity and are thus like knop yarns (described above). Substantial further details were given by Gong and Wright (2002: 47–50). Chenille yarn consists of a cut pile of fibres spaced helically around and secured within a yarn. These yarns were made originally by weaving using a leno structure (where adjacent warp yarns twist around each other and hold weft yarns in place), with the final woven piece being cut into warp-ways strips and used as weft in traditional furnishing textiles. Further explanation was given by Gong and Wright (2002: 55–59). Bouclé yarns belong to a group which includes gimp and loop yarns (both described above). In each case, a component yarn forms tight loops that project from the surface of a central yarn, described by Tubbs and Daniels as 'wavy projections on the surface' (Tubbs and Daniels, 1991: 109). Differentiating between the three types, Tubbs and Daniels stated that 'bouclé yarns exhibit an irregular pattern of semi-circular loops and sigmoid spirals, gimp yarns display fairly regular semi-circular projections and loop yarns have well-formed circular loops' (Tubbs and Daniels, 1991: 109–110). Further description can be found in Gong and Wright (2002: 38–40). Cloud (or grandrelle) yarns consist of two yarns each of a different colour, combined in such a way that each forms the base and cover at some stage to cloud or hide the other thus giving the impression of occasional colour change (Tubbs and Daniels, 1991: 110). Gong and Wright gave further details (2002: 47). An eccentric yarn is a type of gimp yarn (described above).

Tubbs and Daniels (1991: 110) commented: 'Generally, it is produced by binding an irregular yarn such as a stripe or slub … to create graduated half circular loops along the compound yarn'. Further description can be found in Gong and Wright (2002: 38–40). A diamond yarn consists of a thick core yarn (or, possibly, a roving) with two yarns (one S-twisted and the other Z-twisted) cabled around it. Further details were given by Gong and Wright (2002: 36–37).

Fleck yarn is a mixture yarn of 'spotted and short streaky appearance', due to the introduction of small quantities of fibres of different colour or lustre (Tubbs and Daniels, 1991: 110), so it has close similarities to the knop or nepp yarn (see Gong and Wright, 2002: 50–51) mentioned above. Should the reader wish to explore the subject of fancy yarns more, it is worth referring to Gong and Wright (2002), which still stands as an important text nearly twenty years after its publication. Often yarns are illustrated in textbooks, and this is of value to specialists, but the truth is that this is of no real help or assistance to non-specialists. Familiarity is best developed through use and practice.

3.6   SUMMARY

This chapter was concerned with identifying and explaining the main stages in manufacturing the twisted structures known as yarns, using a series of machine types. Initially, with most natural fibres, blending and removal of non-fibrous matter as well as extra-short and entangled fibres are shared objectives. Further to this, the creation of a sliver (an untwisted collection of fibres in rope form) is invariably the intention. After attenuation, the sliver is twisted to produce the desired structure known as a yarn. By the early-twenty-first century, machines common to most fibre-processing systems included: the carding engine, the drawing frame, the comb, the speed (or roving) frame, the spinning frame and the winder. At the same time, the most common spinning system was ring spinning, though it was recognised that if the product range could be extended, it was likely that rotor spinning would take a more dominant position worldwide, as its adoption would eliminate both the speed frame and the winder. In years past, machinery such as ring spinning was focused initially on the requirements of cotton-yarn manufacture, but at first only a coarse product (or count) range was applicable. Both the applicability of the machine to other fibre types as well as the broadening of the count range came about through further technological innovations (a process that appears to be common across all textile innovations). So, it seems likely that this form of innovative attention will be focused in the future on rotor spinning and the extension of its product applicability.

REFERENCES

Gong, R. H., and R. M. Wright (2002), *Fancy Yarns. Their Manufacture and Application*, Cambridge, UK: Woodhead.
McKenna, H. A., J. W. S. Hearle and N. O'Hear (2004), *Handbook of Fibre Rope Technology*, Cambridge, UK: Woodhead.

Tausif, M., T. Cassidy and I. Butcher (2018), 'Yarn and Thread Manufacturing Methods for High-Performance Apparel', in J. McLoughlin and T. Sabir (eds.), *High-Performance Apparel: Materials, Development, and Applications*, Cambridge, UK: Woodhead Publishing, pp. 33–73.

Thompson, R., and M. Thompson (2014), *Manufacturing Processes for Textile and Fashion Design Professionals*, London: Thames and Hudson.

Tubbs, M. C., and P. N. Daniels (1991), *Textile Terms and Definitions*, ninth edition, Manchester: The Textile Institute.

4 Weaving

4.1 INTRODUCTION

Weaving and knitting appear to be universal fabrication processes. The former, and its various developments, both products and processes, are the concern of this chapter. Most woven textiles are biaxial (involving two sets of yarns at 90 degrees). Although there are other possible varieties involving yarns in several directions (known as multi-axial cloths), these are not the concern here, and attention is focused on conventional woven textiles which are biaxial, involving two sets of yarns confronting each other at right angles. At the time of writing, highly specialised and complex machinery had been developed to produce tri-axial cloths (involving the interlacement of two sets of warp yarns and one set of weft yarns, at 60 degrees to each other); this gave a cloth with excellent bursting, tearing and abrasion resistance, and with great stability, making it suitable for use in sail cloths and composites of various kinds.

Basic woven structures (with an interlacement of two sets of yarns, one running at 90 degrees to the other) are introduced, and innovations in weaving are identified and explained briefly. Probably the best-known and most frequently used treatises dealing with varieties of woven structures are Watson (1912 and 1913) and Oelsner (1952), both with numerous subsequent editions. Well-known woven-textile types are introduced in this present book, and various point-paper diagrams (like graph paper though arranged in an eight-by-eight format), which appear to be the best means of representing woven structures, are presented as well. A particularly good introduction was provided by Shenton (2014).

In order to prepare yarns for weaving, it is necessary to gather these into a sheet known as a warp. Relevant machinery, referred to simply as warping machines, gather yarns in parallel format and wind these on to a tubular beam known as a warp beam. The warp yarns from this beam are coated with a size (often a water-soluble, starch-type substance) which allows the yarns to withstand the rigours of the weaving process. After the size has dried, the yarns of the warp are then drawn through the healds (sometimes referred to as heddles) and reed of the weaving frame (known as a loom). This horizontal sheet of warp yarns is separated into two sheets, forming a 'V' shape (in an action known as shedding), and a weft yarn is placed between these two sheets and beaten up into the cloth (which is wound on to a cloth beam at the front of the loom). Subsequently, after weaving, size is removed from the cloth.

4.2 TECHNIQUES AND VARIATIONS

Conventional weaving involves the manipulation and interlacement of two sets of yarns: one set considered to orientate vertically is known as the warp yarns or ends, and the other set considered to orientate horizontally is known as the weft yarns or picks. The interlacement of the two sets of yarns is facilitated through an apparatus known as a loom (in its most basic form this consists of four lengths of wood forming a rectangular frame, with yarn simply wrapped between two opposite sides of this frame). An early innovation was a device known as a shuttle, a wooden projectile which held a pirn (or bobbin) of weft yarn, and this was used to pass a trail of yarn between warp yarns (separated into two sheets). Automation, in many ways associated with the Industrial Revolution, resulted from the past tendency of technological innovations to focus on the use of machinery to replace labour. This focus was largely due to the belief that the costs of labour were high. So, when a process is fully automated, all component actions and transfers are carried out by machines. A semi-automated scenario involves the use of both machines and labour.

By the early-twenty-first century, automated or semi-automated mechanical looms, both attempts to lower the involvement of labour and to reduce total costs, had largely replaced handlooms worldwide in the production of mass-consumed textiles. However, it should be noted that, by that period, handlooms were used frequently in design development within modern industrial settings, often to produce prototype cloth designs. Despite numerous technological refinements in loom structure and the incorporation of many digital innovations, the basic operating sequences associated with weaving remained the same: shed formation, weft-yarn insertion and beating up, all achieved using the loom. Not surprisingly, different approaches to achieving the three actions have involved the introduction of numerous innovations over the years.

The earliest looms, known as ground-horizontal looms, are associated with pre-dynastic times in ancient Egypt (before 3125 BCE) (Hann and Thomas, 2005: 29). Later, perhaps around 1570 BCE, vertical looms (considered by many to be pre-cursors to vertical tapestry looms of the modern era) were introduced into ancient Egypt (Hann and Thomas, 2005: 30). An arrangement known as a warp-weighted loom seems to have been used in ancient Greece around the middle of the first millennium BCE, and possibly before in Hungary (Hann and Thomas, 2005: 32). Various alternatives, such as the back-strap loom, evolved outside Europe. Most innovations in weaving from the nineteenth century onwards were based, however, on what became known as the raised horizontal loom, used throughout much of Europe, Asia and Africa by the late-nineteenth century, though first introduced several hundred years earlier. Important weaving innovations included John Kay's flying shuttle of 1733 CE (Hann and Thomas, 2005: 38) and the Jacquard patterning mechanism introduced by Joseph Marie Jacquard (1752–1834) in 1801. The former dramatically speeded up production and the latter revolutionised the production of complex figured textiles (largely by cutting down on the time and skill required for their production). Hann and Thomas gave more details (2005: 27–41). Further important developments included the introduction of the first power loom and various automatic attachments in the nineteenth century. Major developments in the twentieth century came mainly in the form of what became known as shuttleless looms, where weft yarns were inserted into the cloth by various unconventional means.

   The simplest woven-textile structure, known commonly as plain weave, consists of each warp yarn passing over and under successive weft yarns. A checker-board form is thus produced, with black representing warp yarns raised above white weft yarns; in order to achieve this, it is necessary to create two sheets of warp yarns by raising alternate warp yarns (at say 30 degrees to each other) and to pass a weft yarn between these two sheets (or the shed created between the two). The separation of alternate warp yarns is achieved on relatively modern looms by ensuring that each warp yarn is attached in some way to a heald, a length of wire held in a frame known as a harness, and each alternate warp yarn associated with a heald in one harness of two. Sheds to allow the creation of simple plain weave, and the insertion of weft yarn within, are produced by raising and lowering alternate harnesses. Weft yarn is then inserted within this shed, using the device known as a shuttle (mentioned earlier) which lays a trail of weft yarn within the shed and this weft yarn is then pushed or beaten into the cloth. The basic weaving actions (mentioned earlier) are therefore: shed formation (creating two sheets of parallel yarns), weft-yarn insertion (between the sheets or shed) and beating up of the inserted weft yarn into the body of the cloth. Together, these constitute the weaving cycle. While they are the three basic or primary motions associated with weaving, there are also three secondary motions, known as let-off, take-up and weft selection (and replenishment). The term 'let-off' refers to a motion, which ensures a constant delivery of warp yarn at a suitable tension to the loom. This is generally from a flanged roller known as a warp beam prepared in advance in a process known as warping. As noted previously (in section 4.1), warp yarns are prepared for the stresses and strains of weaving by coating each in a size (or starch), which, on drying, improves their flexibility and strength as well as their abrasion resistance. The second secondary motion, take up, ensures that cloth woven is wound up at a constant pace (generally, this is on a cloth beam, a flanged roller located at the front or output end of the loom). Weft yarns are wound on to small bobbins known as pirns and these are held within shuttles, each of which will travel from one side of the loom to the other side, laying a trail of weft yarn within the shed. Shuttles (each often holding yarn of a different colour) are held in shuttle boxes to the side of the loom and are selected when required. As weft yarns run out, and pirns become empty, it is necessary to replenish these and this constitutes the third secondary motion, weft selection (and replenishment). Various levels of automation have sought to address this issue, with a desire either to refill the shuttle with a pre-wound pirn or simply to replace the shuttle (holding a fully wound pirn within). Various auxiliary motions or additions have developed over the years, most focused on minimising the time when the loom is not in use and some designed to ensure the protection of the cloth produced. Among these are warp-stop motions, weft-stop motions, warp protectors and cloth temples. Warp-stop motions and/or weft-stop motions ensure that the loom stops when a warp or weft yarn breaks. Warp protectors ensure that the loom only stops at a point of the weaving cycle when the shuttle is not within the shed (as much damage could result otherwise). Cloth temples hold the cloth woven to the required width prior to it being wound on to the cloth beam. Some of the innovative focus during the twentieth century was on increasing the speed of the weaving cycle for conventional power looms. The degree of automation was increased, generally through adopting warp- and weft-stop motions, thus ensuring that machine downtime was kept to a minimum. Shuttle looms have various disadvantages. Among these are the great amount of energy required by weft insertion; the high noise levels and vibrations; the great strain on both warp and weft yarns during the weaving cycle; and the difficulty in controlling the movement of the shuttle and ensuring that it does not escape from its intended route.

Alternative means of weft insertion have also been the focus of innovation and various so-called shuttleless varieties of weaving appeared. These offered various advantages over conventional looms, including higher machine productivity, due principally to higher speed and greater loom widths compared to conventional looms. Higher labour productivity and reduced labour costs, due largely to a higher allocation of looms to each worker, were further features. With the elimination of the shuttle, pirn winding and shuttle preparation were eliminated also, and a better working environment, due principally to lower noise levels, became apparent. Up to the early-twenty-first century, four main types of shuttleless weaving were the subject of much research and development: projectile, rapier, jet and multiphase. The operational aspects of each are outlined briefly below.

With projectile varieties (referred to, sometimes, as missile looms), weft insertion is generally from one side only and via projectiles, each with a small clamp or gripper, which grabs the weft yarn from a large package and inserts it into a (narrow) shed. Projectiles are inserted one after the other, once the previous weft has been inserted and beating up has taken place; the number of projectiles held will depend on the cloth width and speed of the weaving cycle. Projectile looms were first introduced in the mid-twentieth century and were subject to continual additions and innovations up to the early-twenty-first century.

Rapier looms are of various types, with earlier models using a single rod (known as a rapier) to carry a trail of the weft yarn from one side of the loom to the other. An advancement was the use of two rapiers, one at each side of the loom, with one rapier taking the yarn halfway into the shed, and transferring it to the other rapier, which carried the yarn to the other side of the shed. The introduction of flexible rapiers minimised the amount of space taken up by each loom. Rapier looms are much faster than shuttle looms but slower than projectile looms.

Jet weaving is of two distinct types (water jet and air jet). With water-jet weaving, a pre-measured length of weft yarn is carried through the shed by a jet of water. Obviously, it is best to use warp and weft yarns which do not readily absorb water. Various filament yarns such as nylon, polyester and glass are well suited. Air-jet weaving uses a jet of air to propel the weft yarn through the open shed.

Innovations relating to a system known as multiphase weaving were probably attempts to overcome speed restrictions imposed by conventional weaving, in particular the necessity to form a single shed and to propel a yarn from one side of this to the other, before opening another fresh shed. In the multiphase case, many different sheds are formed across the machine in a wave formation, so a number of weft yarns can be inserted, one after the other. Therefore, as a section of the shed opens, the weft passes through and the shed closes afterwards. Weft insertion may be via air jets or conventional shuttles. In conventional weaving, a selvedge is created at the right-hand and left-hand extremes of the cloth where the weft yarn in the shuttle is stopped and projected in the opposite direction. Selvedges (often no more than 1.5 centimetres in width) are generally produced in plain weave as this offers the maximum stability and stops the cloth from unravelling. Occasionally, the selvedge will hold a brand name, will indicate the cloth's fibre content or (in textile printing) the colours used. With shuttleless varieties of loom, often secure edges need to be created on each side of the cloth.

Another weaving innovation worthy of comment is circular weaving. With circular weaving, a shuttle enters a shed formed around a machine. This form of weaving is possible on a loom suited mainly to making tubular cloths rather than conventional flat-woven cloths. By the early-twenty-first century, some circular looms could use a (multiphase) shed which took up a wave formation and used up to ten shuttles (driven from beneath by electromagnets), with warp yarns rising and falling as each shuttle passed.

Two further weaving additions are worth mentioning: swivels and lappets. Watson noted that the term 'swivel' was formerly applied to a loom used to weave several narrow-width cloth bands or tapes side by side (Watson, 1925: 334). During much of the twentieth century, the term was used frequently to refer to a weaving process, which depended on the use of several miniature shuttles, known as swivels, positioned across a loom width. These swivel shuttles worked in association with a conventional shuttle (this latter producing a conventional foundation cloth involving one set of warp yarns and one set of weft yarns only) with the swivel shuttles producing figuring (on the foundation cloth) using extra yarns (Watson, 1925: 334). Further details were provided by Watson (1925: 334–362).

Lappet-woven cloths use figuring or whip yarns as extra yarns on the surface of a foundation cloth. Where necessary, these extra yarns traverse in a weft-ways or horizontal direction, after insertion. Watson noted that generally a plain-woven foundation cloth was probably best as this would provide the necessary firm base needed to withstand the sideways pull of the figuring or whip yarns (Watson, 1925: 296). Extensive details were provided by Watson (1925: 258–296).

Tablet weaving is a widely used technique applied to the production of narrow-width or embellished narrow cloths used as belts or as attached borders to more substantial cloths. Traditionally, this hand-based technique was particularly widespread geographically with use recorded in Japan, ancient China, Central Asia, India and the Himalayan countries, Persia, Indonesia, the Caucasus, Syria, Palestine, Egypt, North Africa, Turkey, Greece, Macedonia, Bosnia, Russia, Sweden, Norway, Iceland and France (Schuette, 1956: 9). A worthwhile explanatory monograph is available in the *CIBA Review* series (Schuette, 1956).

A final technique to be mentioned is bleaching, which may be carried out post weaving (often after a cleaning process known as scouring). This involves the removal of all colouring matter. Cloths are bleached (as well as rinsed and dried) to ensure that dye uptake is even when the cloth is immersed in a dye bath.

4.3   PRINCIPLES AND STRUCTURES

As noted above, essentially the component parts of a woven cloth are the warp yarns and weft yarns, interlaced at right angles. In most woven textiles, the order of this interlacement will be on a regular repeating basis, across the whole cloth, with the smallest repeating unit known simply as a repeat. Numerous repeats or interlacing sequences are possible, each referred to as a weave or a woven structure. The intention in this section is to identify and describe briefly the more important of these. Only one category is considered: relatively simple structures with only one set of warp yarns, which are interlaced with one set of weft yarns. For structures that are more complicated and known as compound structures which, in addition to the one set each of warp and weft yarns, also have additional warp and/or weft yarns, the reader is advised to refer to Watson (1925: 24–88). In the design of woven cloths, it is necessary to plan the order of interlacements in advance. The most common system of notation for recording the order of interlacements involves the use of squared paper (known as point paper), ruled with equidistant vertical and horizontal lines in a sequence, or grid, of small, equal-sized squares. Each square represents a point where a warp or weft yarn may cross. Although there is slight variation worldwide, commonly, if the square has a mark within it, this indicates that a warp yarn comes to the surface or face of the cloth (Watson, 1954: 2). Alternatively, where a square has been left blank this represents a weft yarn on the face of the cloth. Where a vertical column of adjacent squares has been marked, this represents a warp float (where the warp yarn remains on the cloth's surface), and where a horizontal row of adjacent squares has been left blank, this represents a weft float (where the weft yarn remains on the cloth's surface). Invariably, at least within the context of the British Isles, the squares on weaving point paper are arranged in blocks of eight by eight, separated by thick lines, which help with easy counting of yarns when planning a design.

As noted in section 4.2, plain weave is the simplest woven structure. It is also the most used. Other common structures are twill weaves and satins (these are warp faced and are known as sateens when viewed on the reverse side, where they will be weft dominant). Each of these common weaves has numerous alternatives, and it is often claimed that all woven structures were derived from plain weave, twills, satins and sateens.

In plain (or tabby) weave, both warp and weft yarns follow a sequence of over one and under one. Plain weave is the firmest cloth type, with maximum possible interlacing built in (Moore, 1998a). Often, yarns of the same weight and density are used in both warp and weft; where this is the case, the resultant cloth is considered to have a balanced sett. A related group, known as hopsack (or basket) weaves, is based on a plain-woven structure. These may be regular or irregular. In the regular variety, an equal number of warp yarns and weft yarns interlace, often producing a balanced sett. Meanwhile, with irregular hopsacks, unequal numbers of warp and weft yarns interlace.

A further related category is warp and weft cords, in which interlacement is with single warp yarns over a given number of weft yarns, with the simplest case being a warp yarn that is raised over two successive weft yarns, and then goes beneath the next two weft yarns, and with the adjacent warp yarn passing beneath two weft yarns and over the next two weft yarns; there is the possibility also of irregular varieties. Equivalent weft-cord varieties are possible as well, together with relevant irregular types.

The fundamental characteristic of pile cloths is the addition of loops or tufts of fibres or yarns. Various forms are possible including velvets, terry towelling, tufted carpets and several other varieties. The production of these can involve weaving, knitting or other techniques. Sometimes piles are formed by loops which are left uncut and sometimes these are cut. Loops may be formed on both sides of a cloth (as is the case with terry towelling) or between two woven cloths (as is the case with a face-to-face-woven cloth, where two parallel cloths are woven with an uncut pile between), or they may be hand knotted or machine tufted (as is the case with carpets). Cloths with looped or pile surfaces are an important textile category and are therefore mentioned a few times in this present book.

Twills and related weaves are a further class of woven structures. These have parallel diagonal effects on the cloth's surface, with the steepness of these depending on the underlying woven structure. The importance of these weaves was stressed by Moore (1998b), who observed that 'the largest, most significant and most versatile of all the weave families are [sic] known as twills'. In a twill weave, all warp yarns will have an identical order of interlacement, but each successive warp yarn across the width of the structure's repeat will start on a different weft yarn. The starting point for adjacent warp yarns will therefore be different. The notation 2/1 (read 'two up and one down') indicates that a warp yarn remains on the surface across two weft yarns and then goes below the surface for the next weft yarn. Numerous alternative twills are of course possible. The steepness of the twill line can of course be changed. A brief explanation was given by Oelsner (1952: 65). Reversed twills are another possibility. A well-known class of these is herringbone weaves, obtained by combining Z and S twills. Diamond-shaped effects are possible also.

Satin and sateen weaves are warp faced and weft faced, respectively (Tubbs and Daniels, 1991: 204). That is, satin weaves have a dominance of warp floats on the surface and sateen weaves, a dominance of weft floats on the surface, though one can be regarded as the opposite of the other, with the reverse side of satins showing sateen structures and the reverse side of sateens showing satin structures. Each weave may be further classed as regular or irregular. With regular satins and sateens, the points of interlacement are distributed in a regular stepping arrangement, and with irregular types the stepping arrangement is at random. The stepping arrangement indicates the step or move number which equals the number of weft yarns crossed by warp-yarn floats in satins and the number of warp yarns crossed by weft-yarn floats in sateens (Hann and Thomas, 2005: 14). If, for example, an adjacent warp thread interlaces with a weft thread three weft threads later than the first interlacement, the stepping arrangement would be regarded as three. Further details were given by Hann and Thomas (2005: 14–15).

In chapter 1 of *Advanced Textile Design*, Watson deliberated on the possible benefits of adding extra yarns to a cloth (1925: 1). Often, the purpose was to increase the warmth-retaining properties or the weight of the cloth, and equally often it was not possible to achieve this in a single cloth without it resulting in a cloth which was 'somewhat coarse in appearance' (Watson, 1925: 1). Extra yarns may, it seems, be introduced for purely embellishment purposes and on occasions for both embellishment and weight (Watson, 1925: 1). Watson observed that: 'By interweaving extra weft, or extra warp, or both extra weft and extra warp ... on the underside of a cloth, it is possible to obtain any desired weight combined with the fine surface appearance of a light single fabric' (Watson, 1925: 1). When extra yarns are included in a cloth, it is important to ensure that these yarns are bound securely into the structure. This can be done by including various stitching yarns, but these must remain out of view in the woven cloth. Watson (1925: 3) outlined the convention (still largely upheld in the early-twenty-first century) for placing stitching yarns and making sure that these were hidden. A double cloth consists of two component cloths, employing two sets of warp yarns and two sets of weft yarns, woven simultaneously into the one component cloth (Moore, 2000a). Various classes of double cloths were reviewed by Moore (2000a, b, c and d). Triple cloths, referred to by Watson (1925: 96) as treble cloths, have three sets each of warp and weft yarns. Aspects of triple cloths and their design were discussed by Watson (1925: 96–106).

4.4 PRODUCT TYPES

It has been noted previously (in sections 4.2 and 4.3) that plain weave is the simplest of woven structures. At least until the early-twenty-first century, it continued to be the most commonly used, as it was inexpensive to produce, compared to other structures with similar component parts, and was durable, with a relatively flat, tight surface, suited to printed designs as well as other forms of finish. An amazingly wide variety of cloth types come under the plain-weave heading with scope for variation in constituent fibres or yarn type, direction and degree of twist in yarns and their density in warp- and weft-ways directions. Among these are georgette (a cotton cloth made in imitation of a silk, with hard-twisted warp and weft yarns); shantung (a coarse silk cloth with slubs, but occasionally of polyester, nylon or viscose); seersucker (often of cotton) with, for example, sections of equal width in the warp direction held at tension during weaving (using two warp beams) thus inducing a puckered section; sheeting (a medium- to heavy-weight cotton used often for bedding); lawn (a light, thin cloth slightly stiff and crease resistant, used often as dress cloth); muslin (light- to medium-weight unbleached cotton, used often in sample garments); poplin (a medium-weight cotton with fine warp yarns, used often in petticoats and as nightwear); canvas (a relatively heavy-weight cloth, often of cotton, used for working clothes as well as artists' canvases); organdie (a light-weight, crisp, sheer cotton, used often as dress cloth); cambric (a soft, closely woven cloth with a glazed finish, used often in apparel); chambray (a medium- to heavy-weight cloth, often in coloured stripes or checks, used often in work wear); gingham (a light- to medium-weight cloth, of open texture, often with dyed yarns in both warp and weft directions, used often as household textiles); georgette (a sheer, light-weight cloth, with doubled, highly twisted yarns, both S and Z, in both warp and weft directions); and voile (a sheer, light-weight, cotton or worsted cloth with highly twisted combed

FIGURE 4.1   Nineteenth-century loom, showing cloth woven, with open shed formed by warp yarns. Donghua University (Shanghai). Photograph by Michael Hann.

FIGURE 4.2   Separation of warp yarns to form a shed, through which weft yarn will pass. The University of Leeds.

FIGURE 4.3   Shuttle with flanged bobbin. The University of Leeds.

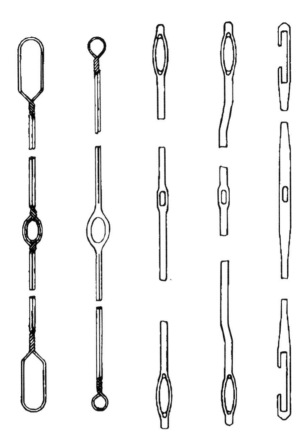

FIGURE 4.4   Various types of heald wires. Re-drawn by JZ.

yarns in both warp and weft directions, used often for women's summer apparel or children's wear).

Both rib weaves and basket weaves are regarded as derivatives of plain weave. Rib weaves are woven cloths consisting of raised ridges in a warp or weft direction. Basket, or hopsack, weave is (as noted previously) a simple modification of plain weave where, instead, two or more warp and weft yarns weave as one. This is best seen as enlarged versions of plain weave with two warp yarns being raised over two weft yarns, or three warp yarns raised over three weft yarns or four warp yarns raised over four weft yarns (known respectively as 2-2, 3-3 or 4-4 basket weaves). Irregular basket weaves are a combination of warp and weft effects. Typical applications for basket weaves include draperies (for example, using heavy-duty cotton yarns with a coarse basket weave) and various apparel uses (especially shirting) in the form of Oxford cloth (with two thin warp yarns interlacing with two soft, much thicker, weft yarns).

Tapestry is invariably based on a plain-woven structure, though it is regarded as a type of weft-faced weave in which the weft threads (often of dyed wool) do not

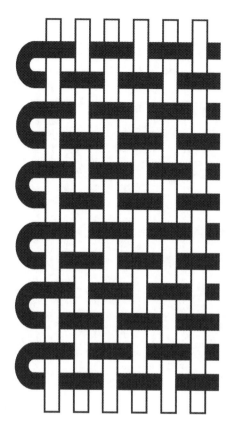

FIGURE 4.5 Conventional plain-woven selvedge. Re-drawn by JW.

necessarily extend across the full width of the cloth, similar to what is referred to as extra-weft figuring. Typically, the warp threads are bleached or natural-coloured cotton or flax yarns (or occasionally, silk) and these are hidden from view in the finished piece by the bulkier weft threads (which interlock with adjacent weft yarns), all within a plain-woven structure. Tapestries are invariably reversible. Although ancient in origin and used in the past to create flat-woven carpets and coverings for walls or furniture, the function in relatively modern times is as woven wall hangings depicting pictorial scenes of various kinds. Tapestry-type techniques were also used in the production of relatively heavy-weight kilims (flat-woven rugs) or in delicate Chinese ke-si woven silks, where the finer details in later-nineteenth-century types were hand painted. (Further details and illustrations can be found in Edwards, 2009.) If weft yarns in different colours are adjacent, vertical slits in the cloth may be created. There are several ways of avoiding this. Probably the most successful solution is plain-woven interlocking tapestry. A similar interlocking effect can be found in many twill-woven Kashmiri shawls.

Pile cloths involve an extra set of warp or weft yarns which form a raised loop pile during the weaving process. These loops may be cut or left uncut. Such cloths were

FIGURE 4.6 Typical selvedge produced during shuttleless weaving. Re-drawn by JW.

used as thermal blankets, as upholstery, towels or, in heavy-duty form, as carpets or rugs. Various pile-type cloths are worth mentioning further. Corduroy is a woven cloth in which a cut-weft-yarn pile forms in a warp direction standing out from the surface of the cloth. Velveteen cloths are occasionally cut to look like corduroy. Woven velvet is a cut-warp-pile cloth, with cut yarns forming across the whole surface. The heaviest pile textile is known as a carpet or rug. In most textile books the words 'carpet' and 'rug' are used interchangeably, though occasionally the word 'carpet' is used where an expanse of covering is expected and the word 'rug' where a smaller area is intended. So, the word 'rug' is used in this present book to refer to a floor covering which measures no more than 1.25 metres by 2.0 metres (this appears to be a common size for many hand-tufted rugs) and the word 'carpet' applied where the article is larger than this. The word 'tufted' refers to a heavy-duty surface pile of the type used in carpets and rugs. Tufts may be inserted by hand or machine into a preformed carpet backing or can be inserted during the weaving process between one or more inserted weft yarns. Often books refer to various carpet types: Wilton, Axminster, machine-tufted, hand-tufted and kilim are the most common. Wilton pile carpets are machine woven using a Jacquard loom. The product can be either

FIGURE 4.7 Shuttleless weaving selvedge with additional support (shown on left). Re-drawn by JW.

highly patterned and multi-coloured or plain with no surface design. After weaving, the pile may either be left looped or be cut. Axminster pile carpets may be of various types depending upon how the tufts are inserted into the structure. Machine-tufted carpets use a pre-made substrate to receive pile yarn inserted using tufting needles. Machine-tufted carpets may be of a single colour or may be multi-coloured. Hand-tufted carpets are where each dyed tuft is inserted by hand between one or more weft yarns (forming a plain-weave structure) during the weaving process. This may involve numerous colours, and the pattern may be exceedingly complex. Kilim (sometimes spelt gelim, kelim or khlim) are hand-woven floor coverings without a pile, in plain weave, with dyed wool weft yarns which hide undyed cotton warp yarns. Structurally, kilims are like tapestry products where weft is discontinuous but, in the case of kilims, adjacent weft yarns are not interlocked with each other so vertical slits develop.

A gul (or gol) is the most common motif class in hand-tufted rugs and carpets. This is a polygonal motif associated with Turkoman rugs and carpets. Invariably the motif is based on an octagon or hexagon and exhibits bilateral reflection symmetry

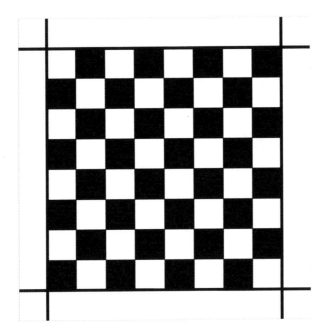

FIGURE 4.8   Plain weave. Drawn by JZ.

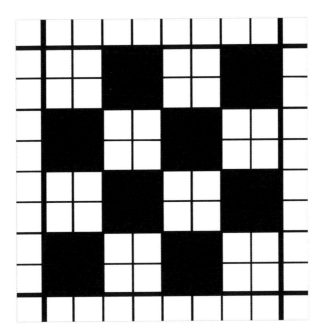

FIGURE 4.9   Regular hop sack (2 × 2). Drawn by JZ.

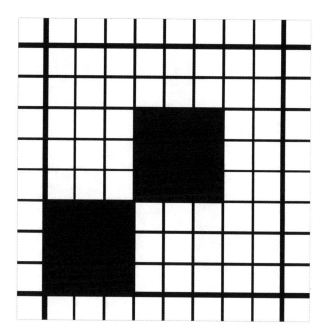

FIGURE 4.10   Regular hopsack (3 × 3). Drawn by JZ.

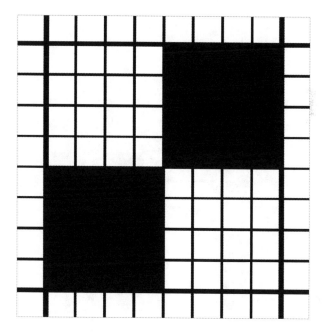

FIGURE 4.11   Regular hopsack (4 × 4). Drawn by JZ.

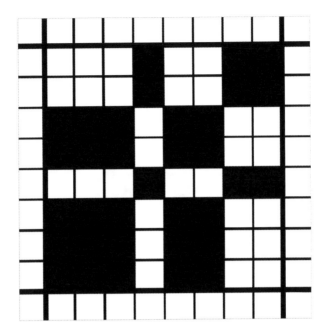

FIGURE 4.12   Irregular hopsack. Drawn by JZ.

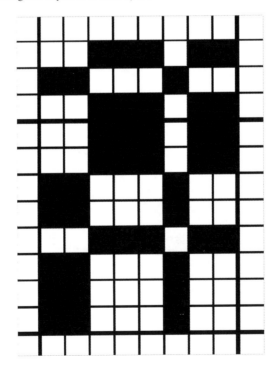

FIGURE 4.13   Irregular hopsack. Drawn by JZ.

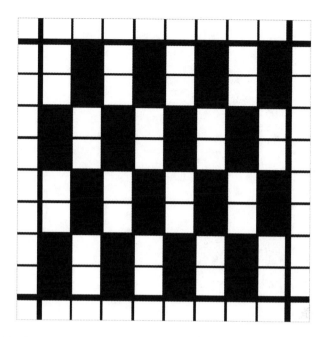

FIGURE 4.14 Regular warp cord (over 2). Drawn by JZ.

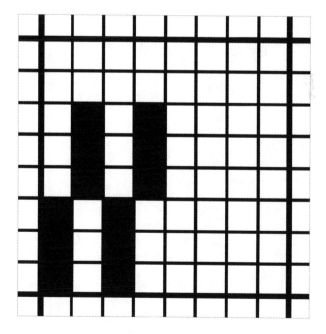

FIGURE 4.15 Regular warp cord (over 3). Drawn by JZ.

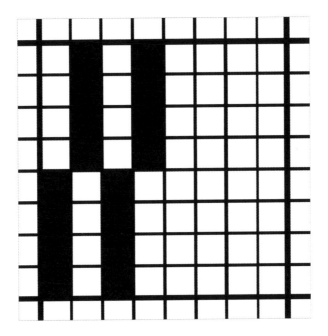

FIGURE 4.16   Regular warp cord (over 4). Drawn by JZ.

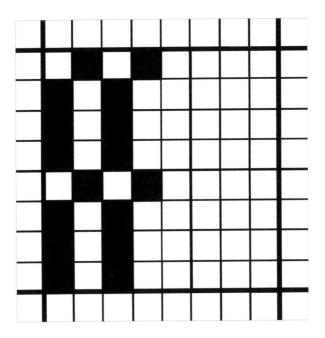

FIGURE 4.17   Irregular warp cord. Drawn by JZ.

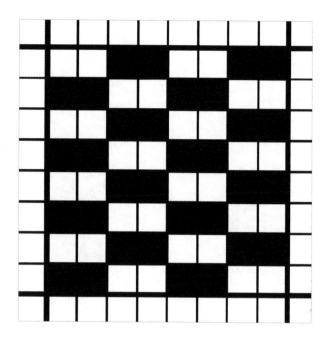

FIGURE 4.18 Regular weft cord (over 2). Drawn by JZ.

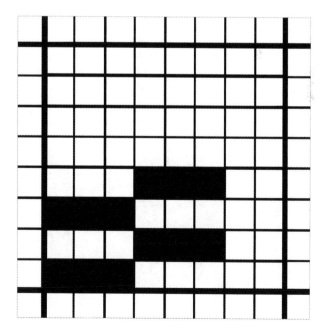

FIGURE 4.19 Regular weft cord (over 3). Drawn by JZ.

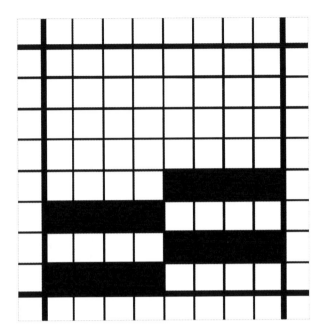

FIGURE 4.20   Regular weft cord (over 4). Drawn by JZ.

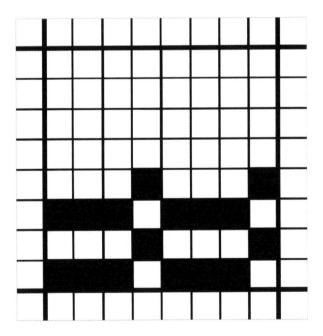

FIGURE 4.21   Irregular weft cord. Drawn by JZ.

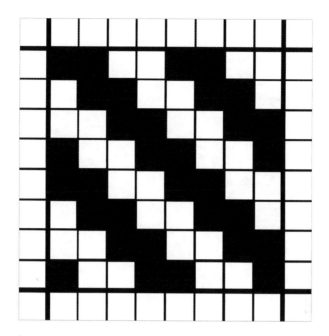

FIGURE 4.22   2/2 twill (S). Drawn by JZ.

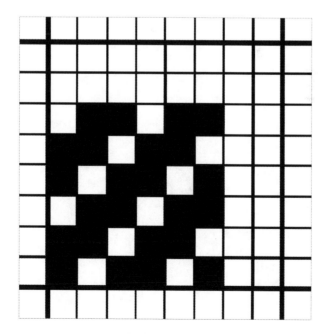

FIGURE 4.23   2/1 twill (Z). Drawn by JZ.

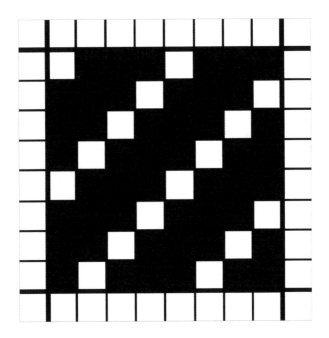

FIGURE 4.24 3/1 twill (Z). Drawn by JZ.

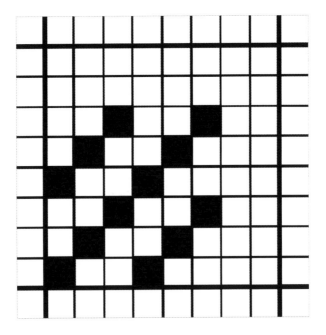

FIGURE 4.25 1/2 weft-faced twill. Drawn by JZ.

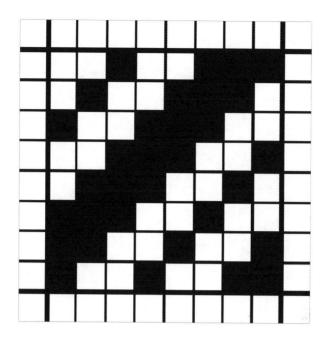

FIGURE 4.26 3/2/1/2 twill (Z). Drawn by JZ.

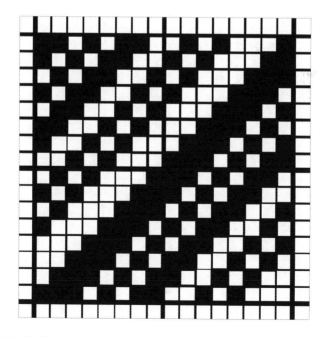

FIGURE 4.27 Twill on 16 × 16. Drawn by JZ.

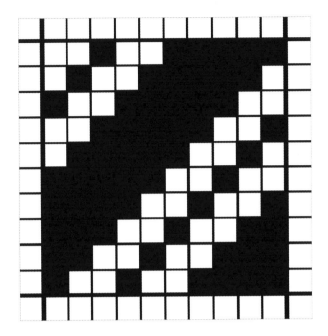

FIGURE 4.28   Twill on 10 × 10. Drawn by JZ.

FIGURE 4.29   Steep twill. Drawn by JZ.

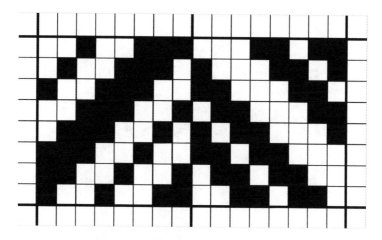

FIGURE 4.30 Herringbone twill from 3/2/1/2 twill. Drawn by JZ.

FIGURE 4.31 Diamond twill developed from 2/2 twill. Drawn by JZ.

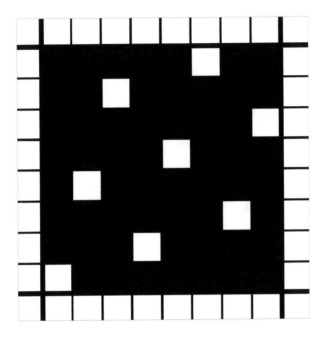

FIGURE 4.32   Satin weave on 8-end repeat. Drawn by JZ.

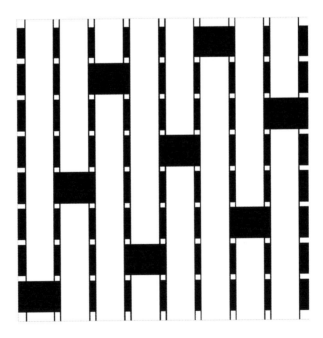

FIGURE 4.33   Plan view of satin weave. Drawn by JZ.

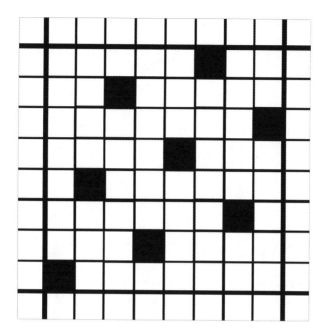

FIGURE 4.34   Sateen weave. Drawn by JZ.

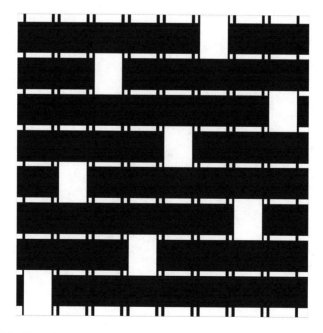

FIGURE 4.35   Plan view of sateen weave. Drawn by JZ.

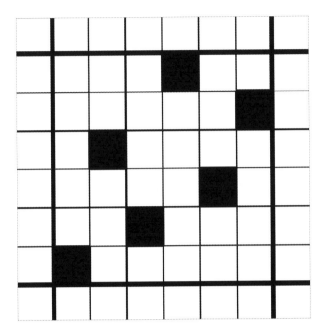

FIGURE 4.36 Random distribution of interlacement points (6 × 6). Drawn by JZ.

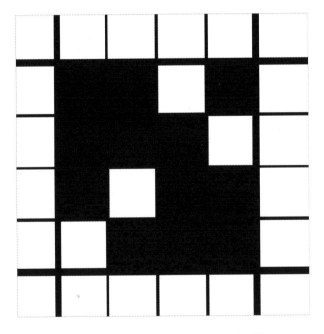

FIGURE 4.37 Random interlacement points (4 × 4). Drawn by JZ.

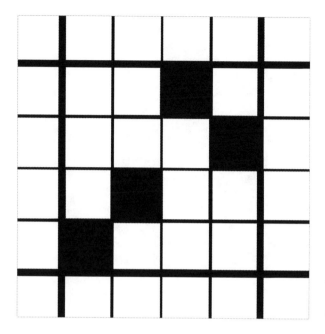

FIGURE 4.38   Random interlacement points (4 × 4). Drawn by JZ.

FIGURE 4.39   Diamond diaper weave. Drawn by JZ.

FIGURE 4.40   Large reversing twill. Drawn by JZ.

FIGURE 4.41   Diaper weave using two basic weaves. Drawn by JZ.

FIGURE 4.42   Diaper weave based on sateen weave. Drawn by JZ.

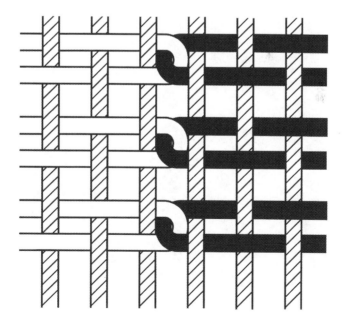

FIGURE 4.43   Interlocking, plain-woven tapestry weave. Drawn by JW.

FIGURE 4.44 Late-nineteenth-century Qing-dynasty tapestry. Held at the University of Leeds.

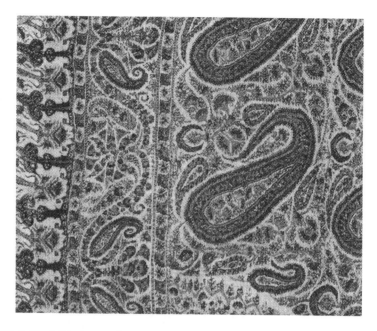

FIGURE 4.45 Mid-nineteenth-century Kashmir (India) shawl. Held at the University of Leeds.

FIGURE 4.46 Central Asian carpet gul. Re-drawn by JW.

FIGURE 4.47   Central Asian carpet gul. Re-drawn by JW.

FIGURE 4.48   Central Asian carpet gul. Re-drawn by JW.

(where the right-hand side appears to reflect the left-hand side). Sometimes the motif is repeated in the form of an all-over regular repeating pattern in the main field of the design, and sometimes it is presented in border form as a regularly repeating component. Even though Stone's treatise was produced over two decades ago, it still stands as an important compendium of carpet/rug types and their component parts (Stone, 1997).

As noted previously (section 4.3), satin weaves are warp faced, whereas sateen weaves are weft faced (though one converts to the other on simply turning the cloth from face side to reverse side). A typical satin structure may involve each warp

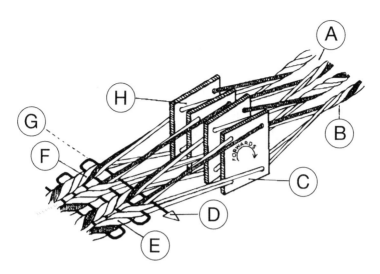

FIGURE 4.49 Schematic illustration of a tablet weaving loom, where the shed of warp yarns is controlled by a series of tablets or cards. A = end of warp; B = twisting of warp yarns; C = a tablet with holes through which warp yarns pass; D = a weft yarn which has passed through the shed; E = right-hand side of cloth; F = left-hand side of cloth; G = the fell of the cloth; H = tablet with holes at left-hand side of loom. Re-drawn by JZ.

yarn floating over four weft yarns. Satins are smooth with great drapability, though the floats snag readily. End uses are as ribbons and trimmings of various kinds. Originally, many satins used silk yarns as warp and cotton yarns as weft, and so were highly lustrous on the face side though relatively dull on the reverse side.

Jacquard designs may contain combinations of plain, twill and satin structures, but often these are expensive to produce. Damask, often from flax or cotton, may contain such a combination of structures. Short floats ensure a cloth is more durable whereas longer floats give a more lustrous cloth. Leno cloths are another important category. In this case, two adjacent warp yarns twist around each other, after each weft yarn has passed through the shed.

4.5 WORLDWIDE DISTRIBUTION OF TECHNIQUES AND PRODUCTS

Probably the most extensive review of textile manufacture, both historically and worldwide, was the *CIBA Review* (with several monographs published each year between the late 1930s and early 1970s, by Ciba Geigy, of Basel, Switzerland). There are, however, numerous books worth consulting, including Weibel (1952) and Wilson (1979) as well as Gillow and Sentance (2004), all of which have covered the nature of traditional textile products and processes worldwide. Numerous publications associated with museums are worthy of reference as well, including the publications from the Victoria and Albert Museum (London) and the British Museum (London).

4.6   SUMMARY

This chapter was concerned with explaining the rudimentary principles of conventional weaving, i.e. creating a woven cloth by interlacing at right angles two sets of yarns (known as warp and weft yarns), invariably using a device known as a loom. By the early-twentieth century, the bulk of industrial woven-textile manufacture worldwide was through either semi-automatic or fully automatic shuttle varieties of loom, where a shuttle containing weft yarn was passed between sheets of warp yarns (which had been separated into two parts). Various shuttleless varieties of loom were developed and used in the late-twentieth century. Four varieties are worth mentioning: projectile, rapier, jet and multiphase systems. Each of these offered potentially great benefits over shuttle varieties of loom and, by the early-twenty-first century, adoption of each was extensive worldwide. As with most textile innovations, the continuous challenge for the machinery manufacturer, with each loom type, was to extend the product applicability of each innovation.

REFERENCES

Edwards, C. (2009), *How to Read Pattern*, London: Herbert Press.
Gillow, J., and B. Sentance (2004), *World Textiles: A Guide to Traditional Techniques*, London: Thames and Hudson.
Hann, M. A., and B. G. Thomas (2005), *Patterns of Culture. Decorative Techniques*, Leeds, UK: The University of Leeds International Textiles Archive.
Moore, R. (1998a), 'Classical Woollen and Worsted Cloth Designs, No. 1: The Plain Weave', *Wool Record*, October, pp. 14–15.
Moore, R. (1998b), 'Classical Woollen and Worsted Cloth Designs, No. 3: The Twills (part 1)', *Wool Record*, December, p. 17.
Moore, R. (2000a), 'Classical Woollen and Worsted Cloth Designs, No. 23: Double Cloth Weaves (part one)', *Wool Record*, August, pp. 34–35.
Moore, R. (2000b), 'Classical Woollen and Worsted Cloth Designs, No. 24: Double Cloth Weaves (part two)', *Wool Record*, September, pp. 84–85.
Moore, R. (2000c), 'Classical Woollen and Worsted Cloth Designs, No. 25: Double Cloth Weaves (part three)', *Wool Record*, October, pp. 44–45.
Moore, R. (2000d), 'Classical Woollen and Worsted Cloth Designs, No. 26: Double Cloth Weaves (part four)', *Wool Record*, November, pp. 30–31.
Oelsner, G. H. (1952), *A Handbook of Weaves*, New York: Dover.
Schuette, M. (1956), *Tablet Weaving*, CIBA Review, no. 117, November.
Shenton, J. (2014), *Woven Textile Design*, London: Laurence King Publishers.
Stone, P. F. (1997), *The Oriental Rug Lexicon*, London: Thames and Hudson.
Tubbs, M. C., and P. N. Daniels (eds.) (1991), *Textile Terms and Definitions*, ninth edition, Manchester: The Textile Institute, previously with different editors.
Watson, W. (1912), *Textile Design & Colour. Elementary Weaves and Figured Fabrics*, London: Longman.
Watson, W. (1913), *Advanced Textile Design*, London: Longman.
Watson, W. (1925), *Advanced Textile Design*, London: Longman.
Watson, W. (1954), *Textile Design and Colour, Elementary Weaves and Figured Fabrics*, sixth edition, London: Longmans.
Weibel, A. C. (1952), *Two Thousand Years of Textiles*, Detroit: Detroit Institute of Arts.
Wilson, K. (1979), *A History of Textiles*, Boulder, CO: Westview Press.

5 Knitting and other forms of yarn manipulation

5.1 INTRODUCTION

In addition to weaving, there are several further means of creating a cloth. Foremost among these is knitting, which brings one or more yarns into association through the formation of loops. Knitting has been defined simply as the process of forming a cloth by 'the intermeshing of loops of yarn' (Tubbs and Daniels, 1991: 164). A few of the machine types used to produce knitted cloth are reviewed in this chapter. Further means of creating cloth from one or more yarns include netting, macramé, lace making, sprang, braiding and crochet; these are explained briefly below, and an outline is given of the embellishment technique known as embroidery.

5.2 KNITTING TECHNIQUE VARIATIONS

Knitted cloth consists of a series of loops, produced using hooked needles (with spring, or bearded, needles and latch needles being the most common at the time of writing). These needles facilitate the creation of loops and the pulling of one loop through the previously produced loop, fundamental actions towards the creation of a knitted cloth. All needles must therefore 'have some method of closing … to retain the new loop and exclude the old loop' (Spencer, 2001: 19). The arrangement of loops is considered in terms of columns and rows. A column of loops is known as a wale and a row of loops as a course. The word 'stitches' is used to refer to intermeshed loops. The density of loops, or stitch density, refers to 'the total number of loops in a measured area … and not to length of yarn in a loop [which is referred to as stitch length]' (Spencer, 2001: 17). Important machinery sub-division is between circular machines and flat machines.

Up to the early-twenty-first century, needle types on machines may have varied, with latch and bearded needles being the most common. From the beginning of the industrial era, the process of loop formation remained the same: yarn was introduced to a hooked needle, which was then closed and moved to pull the introduced yarn through a previously created loop and, in so doing, created a further loop. The process was well explained and illustrated by Power (2015: 290). The two distinct types of knitting techniques are weft knitting, where only one yarn may be used to feed all needles, and warp knitting, where each needle has its own yarn supply and rows of loops are produced simultaneously in an interlocked or zigzag format. All warp-knitted cloths are made in flat format on flat (needle-bar) machines, whereas many weft-knitted cloths are made in tubular format on circular machines and some also in flat format using flat machines. Flat knitting machines can shape or fashion the sides of a knitted cloth, thus creating fully fashioned pieces. Circular knitting is, however, a much faster method of knitting. Although weft knits can theoretically be made from one yarn supply, in practice more than one yarn feed is normal.

As indicated above (in the first paragraph of this section), each needle is associated with a means of opening and closing that allows the needle to take a new yarn and release the previously produced loop. Both Spencer (2001: 20–29) and Mukherjee (2018: 200–201) gave explanations and illustrations of needle types and their operation.

The term 'gauge' is used to refer to the density of needles in a machine (and thus the potential fineness or coarseness of the resultant knitted cloth); this is generally measured in terms of the numbers of needles per English inch. A high gauge, of say 12, 14 or 18, will enable a fine cloth (say for sportswear or lingerie) to be manufactured, whereas a lower gauge, of say 3 or 5, will enable the production of a relatively coarse cloth, for outerwear or similar use, to be manufactured.

The term 'circular knitting machine' generally refers to a weft-knitting machine in which needles have been arranged around the circumference of one or more cylinders (Tubbs and Daniels, 1991: 58). Double-cylinder knitting uses two cylinders, one above the other, and one set of double-ended needles (Tubbs and Daniels, 1991: 91). A dial-and-cylinder machine is a circular knitting machine which uses two sets of needles (one vertical and in parallel around the cylinder of the machine and the other arranged horizontally and radially on a disc, or dial, which fits at the top of the cylindrical arrangement) (Tubbs and Daniels, 1991: 85). A garment-length machine is a machine used specifically for the knitting of individual garment panels (Tubbs and Daniels, 1991: 132).

A crochet-knitting machine is a warp-knitting machine, which uses latch needles (or, occasionally, another type of needle known as a carbine). Because of its use in the production of cloth trimmings of various kinds, it is sometimes referred to as the trimming machine (Tubbs and Daniels, 1991: 77).

An important late-twentieth- and early-twenty-first-century development was complete garment technology where a complete garment (with no further assembly or sewing of panels required) could be produced using one machine only. By the beginning of the third decade of the twenty-first century, important manufacturers included Shima Seiki (with headquarters in Japan) and Stoll (with headquarters in Germany).

5.3 BASIC KNITTED STRUCTURES

Knitted cloth, when compared to equivalent woven cloth, is regarded as highly elastic and, historically, was selected for use in stockings and other items of clothing which required stretch. With high elasticity, knitted cloths invariably give excellent drape, but they are rather thick compared to equivalent woven cloths. Elasticity and other characteristics are of course closely dependent on the fibre content and structure of each cloth. A woven cloth with a high percentage of elastane fibres will be very elastic. Therefore, comparisons are best made with equivalents only and this is the assumption throughout this present book. Knitted cloth will often resist wrinkling but will not take a crease easily. The potential for design variation in knitting is immense and is introduced by different combinations of stitches, fibres, yarns and colours. Elsasser (2005: 139–153) gave a good brief description of the range of techniques and properties.

Over the centuries, hundreds of different knitting stitches have been used. The basic weft-knit cloths are as follows: plain (or single-jersey) knits; rib (or double-jersey) knits; purl knits; interlock knits. Spencer (2001: 60) observed that each of the four cloth types mentioned here consists of a different combination of face- and/or reverse-side stitches, and each may exist on its own, in a slightly modified form, or in combination with one of the others.

Plain, single-jersey knits, also known simply as single knits, can be produced at a fast rate; they are relatively inexpensive but with an immense design potential. They can be made across the full spectrum of weights, from light previously (with different editors) weight to heavy previously (with different editors) weight; although the cloth type is relatively light-weight, the end weight will obviously depend to a large degree on the types of yarns used. They have good stretch properties, especially in the cross-wise direction, but tend to ladder easily if a stitch breaks, and to curl when cut (though special finishes can overcome this curling tendency). Visible flat vertical lines on the front and dominant horizontal ribs on the back are characteristic features. End uses for single knits include dresses, sweaters, T-shirts, hosiery and underwear. Spencer provided further explanation (2001: 61–67). Intarsia is a variety of single-knit cloth, often with different courses using different colours and yarn types producing a highly embellished, multi-coloured effect; the cloth appears the same on the face and on the reverse, and is typically used in blouses, shirts and light-weight sweaters. The potential for highly patterned cloths is offered using a Jacquard mechanism on a circular knitting machine, in conjunction with combinations of stitches, colours and textures. Floats are a characteristic feature of Jacquard knits. Further weft-knitted cloth types, based on single-knit structures, include knitted terry, which incorporates an uncut pile on the reverse side. Although knitted terry is less stable than the woven equivalent, the knitted variety has better drape and is used for towelling, robes and beachwear. Further single-knit cloth types include fleeces and knitted velour. Typically, fleeces are single-jersey cloths with brushed piles, using cotton, cotton/polyester, wool or acrylic fibres. Knitted velour, used commonly as loungewear and sportswear, is weft knitted with a cut and brushed loop pile on the reverse side; although, like woven looped cloths such as velvet, the knitted equivalent has greater stretch and softness. Sliver-pile knits, another single-jersey variety, have a thick sliver pile on the reverse side and commonly function as imitation fur.

Rib knits display vertical columns (or wales) of loops on both front and back. Because the structure is slightly more complicated, rib knits are slower to produce compared to single-jersey knits. Rib knits are often used to provide the ribbing parts at the cuffs, necklines and lower edges of sweaters. A 1 × 1 rib has alternate wale stitches knitted to the front and then the back of cloths. So, ribs can be described as 2 × 2 or 3 × 3, etc. Rib knits are even more elastic than single-jersey knits. The edges do not curl but (like single-jersey knits) ladder easily. Two sets of facing and intersecting needles are required. Rib knits may be created on flat or circular machines, and the resultant cloth may be used for knitted hats and men's hosiery as well as the various sweater uses mentioned previously in this paragraph. Spencer provided further explanation (2001: 67–72). Cardigans, both half-cardigan and full-cardigan varieties, are variations of rib knits, in both cases with a raised effect and a resultant thicker cloth. Half cardigans have reduced stretch in the width-ways direction compared to other rib cloths, are not reversible and, when coarsely knit, are used often for sweaters. Full-cardigan knits are thick and bulky and appear the same on both sides. Often made from wool or acrylic in a coarse gauge in heavy weights, full cardigans are used frequently in fashion end uses. Milano ribs are a further rib variant. Half-Milano rib is an unbalanced structure, often knitted in coarse gauge and used in the making of sweaters. Full Milano is finely knit with good coverage and dimensional stability; uses are mainly as women's suiting cloths.

Purl knits have alternate courses of knit stitches and purl stitches on both sides of the cloth, so appear the same on both sides. Purl cloth does not curl as readily as other weft-knitted cloths. Variations such as 3×1 and 2×2 are possible, and there is much scope for further embellishment, so this type is associated with numerous designs. The structure is often used in sweaters and children's clothing, as well as underwear and sportswear. Spencer provided further explanation (2001: 76–81).

Interlock knits are reversible, with a smooth appearance on each side. These do not curl, are firmer, but less extensible compared to other weft-knitted cloths. This cloth is heavier and thicker than standard rib knits using the same yarn types. Often interlock knits are used for outerwear as well as dresses and skirts, maybe in wool, acrylic or polyester yarns, with cotton and polyester/cotton blends used commonly in underwear. With interlock, two feeds are required to create one course. Interlock stitch is a variation of rib stitch. The front and back are the same, though the cloth is usually heavier and thicker than regular rib knits. Spencer provided further explanation (2001: 73–76).

Weft-knitted cloths have high elasticity, good thermal insulation and resistance to creases, but, in general, they unravel easily (or ladder readily) and retain shape poorly. Overall, warp-knitted cloths are more stable, have good shape retention and are not as prone to unravelling and laddering as weft-knitted cloths, but do not drape as well as weft-knitted cloths.

Double knits are produced on machines with two interlocking sets of needles set at an angle to each other. Typically, resultant cloths (maybe of wool, polyester or blends of the two) are more compact and stable and are thicker and heavier than single-knit cloths using similar yarns. Cable-knitted cloth is double knitted, using a loop-transfer technique, which gives a braid-like appearance; cable cloths are used widely in sweaters. Bird's eye is another double-knitted cloth; often in a multi-coloured effect, with an eyelet (resembling a bird's eye) created on the surface, the cloth is used widely in women's wear. A further variety of double knit is pointelle, which has a lace-type effect created by numerous transferred stitches and is used often in children's-wear cloths.

Warp-knitted cloth consists of vertically oriented loops. Yarns zigzag along the length of the cloth interlocking with adjacent loops. Warp-knitted cloth is made on a special knitting machine with yarns from a warp beam. Unlike weft knits, they are knitted from multiple yarns, with yarns connecting loops in adjacent wales. The face side has slightly inclined vertical knitting loops whereas the reverse side has inclined horizontal floats. They do not unravel easily. Warp-knitted cloths are constructed with yarn loops formed in a vertical or warp direction. All yarns used for a width of a warp-knit cloth are placed parallel to each other in a manner like warp yarns in weaving. These structures have a greater resistance to laddering than equivalent weft-knitted cloths. The two most common warp-knitted cloths are tricot and raschel, with the former being of greater popularity (at least up to the time of writing). Tricot cloths are produced at very high speeds, usually in a plain or simple geometric design and generally from filament yarns (which offer a desirable uniform diameter). Raschel cloths are produced from spun staple or filament yarns of various fibre types and counts. In raschel cloths the wales appear braid-like, as the loops cross. Many raschel cloths offer a surface design, which appears almost three-dimensional, often with the open network shapes associated with lace or crochet. Ray and Blaga (2018: 227–258) gave a comprehensive review of warp-knitted product types, techniques and machines.

Uses for warp-knitted cloth include ladies' underwear, various apparel-related items, such as sportswear lining and inner shoe lining, leisure wear and safety vests, as well as furnishings, laundry baskets, mosquito nets, aquarium fish nets, various automobile uses (including car cushions, head-rest linings, sunshades and linings for motorbike helmets), and across numerous industrial uses such as masks, caps and gloves.

5.4   NETTING, MACRAMÉ, LACE MAKING, SPRANG, BRAIDING AND CROCHET

Nets, macramé, laces, sprang, crochet and braid are each types of cloth created through a series of knots or other manipulation of yarns. Each is explained briefly in this section.

As observed by Emery (1994: 46), the term 'netting' usually 'connotes an open textured single-element fabric with meshes of fixed dimensions secured by knots'. Broadly, the word is used to 'suggest nothing more than an open-meshed' structure (Emery, 1994: 46). In nets the openings may be in the form of equilateral triangles, squares or regular hexagons. Probably the best-known end use for nets is in fishing and as a resistance to mosquitos, though various sporting uses come to mind, including netting used in tennis, soccer, hockey and basketball.

Macramé is produced using knots of various forms. The term and technique are believed to be Arabic in origin. In Europe in the nineteenth century macramé was used as tablecloths, curtains and bed quilts, a range that was extended during the late-twentieth century to include wall hangings and various articles of clothing as well as a wider range of furnishings. By the early-twenty-first century, macramé was regarded widely as a textile craft, with beginner's kits and basic instructions available in most craft shops.

Lace is an open net-type textile worked often in bleached yarn, with varieties produced by hand, often classed into one of three types: embroidered lace; needlepoint lace; bobbin (or pillow) lace. In the early-twenty-first century the term 'lace' was applied to any openwork cloth created (either by hand or by machine) through twisting, looping or knotting of yarns, made from flax, cotton, silk, polyester, rayon, etc., or any combination of these. Invariably, lace includes openwork areas adjacent to areas filled in with yarn, so this is a cloth which is open mesh in parts and opaque in other parts. A good introduction to the subject was given in the late-twentieth century by Pond (1973).

Sprang is an ancient method like netting, but the resultant cloth is made from a series of warp yarns, laid out on a rectangular frame or simply stretched between two parallel sticks. Historical documentation is limited, due largely to the fact that many early examples were identified incorrectly as knitted or lace. Further details of the technique were given by Barber (1991: 122–125) and brief explanation was given by Gillow and Sentance (2004: 56).

Braiding (or plaiting), found in numerous cultures worldwide, is a process of interlacing three or more flexible structures such as yarns 'in such a way that they cross one another in diagonal formation' (Emery, 1994: 35), creating flat, tubular or solid constructions. A braid (or plait) from the interlacement of more than three strands may be the source of more complex designs. Often braids were used in trimmings to garments (especially formal dress and military uniforms), in regalia of various kinds, in sword and other weapon embellishment and, frequently, also, as a form of human hair arrangement. Kyosev (2014) provided a worthwhile review of the textile context and braiding developments during the Industrial Revolution and later.

Crochet is a process of producing a cloth using a hooked implement, known as a crochet hook (of bone, steel or other metal, as well as bamboo, wood or plastic), and employing the following: chain, slip, single-crochet, half-double-crochet or double-crochet stitches. Explanations and details of each were given by Emery, who commented that crochet was basically 'a kind of chaining' (Emery, 1994: 43). It should be noted that English terminology relating to crochet appears to differ from country to country, with a marked difference between what appears to be acceptable in the British Isles and the USA contexts. Hall (1981c) provided a good, understandable review.

5.5   EMBROIDERY

Embroidery is the application of coloured yarn, often to a natural-coloured or bleached, plain-woven linen or cotton ground cloth using one or a combination of embroidery stitches. A particularly popular form of embroidery, using canvas as a base, is known as needlepoint; Hall (1981a) provided a worthwhile review. Emery (1994: 232) defined embroidery as 'the embellishment of ... [cloths] ... by means of needle-worked stitches'. She continued by explaining the nature of the following: flat stitches; running stitches; overcasting; satin stitches; back stitches; stem stitches; double-faced cross stitches; long-armed cross stitches; button-hole stitches; feather stitches; chain stitches; knotted looping; detached stitches; composite stitches; and pile stitches, and gave explanations of canvas work, crewel embroidery, couching and cut work. A review of the main embroidery stitches was provided by Hall (1981b).

FIGURE 5.1 Knitted 1 × 1 rib cloth. Drawn by JZ.

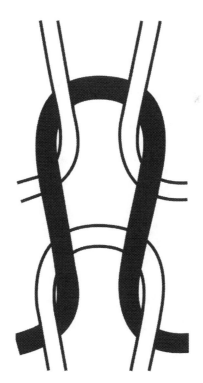

FIGURE 5.2 Example of loop. Drawn by JW.

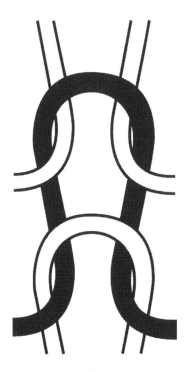

FIGURE 5.3   Example of loop. Drawn by JW.

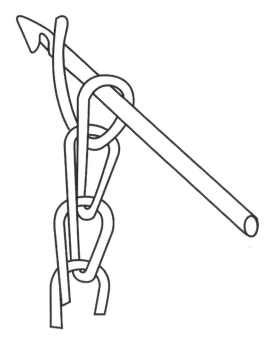

FIGURE 5.4   Hand-crochet loop of similar formation to knitted loop. Re-drawn by JW.

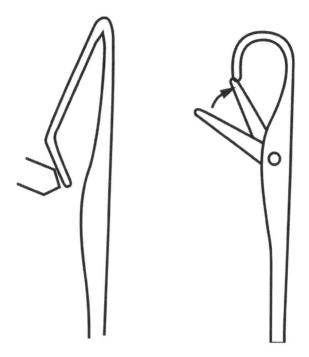

FIGURE 5.5   Bearded or spring needle (left) and latch needle (right). Drawn by JZ.

FIGURE 5.6   Example of lace. The University of Leeds.

FIGURE 5.7   Example of lace. The University of Leeds.

FIGURE 5.8   Nineteenth-century Turkish embroidery. The University of Leeds.

FIGURE 5.9 Nineteenth-century Turkish embroidery. The University of Leeds.

FIGURE 5.10 Nineteenth-century Turkish embroidery. The University of Leeds.

FIGURE 5.11   Nineteenth-century Turkish embroidery. The University of Leeds.

FIGURE 5.12   Indian embroidery showing mirror work, early-twentieth century. The University of Leeds.

FIGURE 5.13   Nineteenth-century Qing-dynasty embroidery. The University of Leeds.

FIGURE 5.14   Nineteenth-century Qing-dynasty embroidery. The University of Leeds.

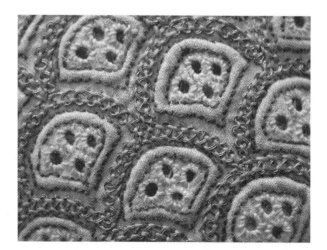

FIGURE 5.15   Use of metallic yarn on nineteenth-century Turkish embroidery. The University of Leeds.

FIGURE 5.16   Nineteenth-century Qing-dynasty embroidery. The University of Leeds.

FIGURE 5.17   Detail of nineteenth-century Qing-dynasty embroidery. The University of Leeds.

5.6 SUMMARY

Knitting is a means of constructing cloth through interlocking loops of yarn formed by means of needles. Vertical columns of loops are known as wales and horizontal rows of loops are known as courses. Compared to their woven equivalents, knitted cloths are regarded as elastic, porous, light-weight, wrinkle resistant and comfortable to wear. Two knitted-cloth types, weft knitted and warp knitted, rely on the use of different machines. An outline of techniques and products was given in this chapter. Attention was turned also to identifying and mentioning briefly other fabrication techniques, including netting, macramé, lace making, sprang, braiding and crochet. An outline was given also of the embellishment technique known as embroidery. There are various further cloth embellishment or construction techniques not mentioned here; foremost among these are probably quilting, patchwork and appliqué. Should the reader wish to explore these subject areas, it is worth referring to the two publications by Walker (1981 and 1983).

REFERENCES

Barber, E. J. W. (1991), *Prehistoric Textiles*, Princeton, NJ: Princeton University Press.

Elsasser, V. H. (2005), *Textiles. Concepts and Principles*, second edition, New York: Fairchild.

Emery, I. (1994), *The Primary Structures of Fabrics*, London: Thames and Hudson in association with the Textile Museum (Washington D.C.), previously 1966 and 1980.

Gillow, J., and B. Sentance (2004), *World Textiles: A Visual Guide to Traditional Techniques*, London: Thames and Hudson.

Hall, D. (1981a), *Needlepoint*, in the Good Housekeeping series, London: Ebury Press.

Hall, D. (1981b), *Embroidery*, in the Good Housekeeping series, London: Ebury Press.

Hall, D. (1981c), *Crochet*, in the Good Housekeeping series, London: Ebury Press.

Kyosev, Y. (2014), *Braiding Technology for Textiles*, Cambridge: Woodhead.

Mukherjee, S. (2018), 'Weft-Knitted Fabrics', in T. Cassidy and P. Goswami (eds.), *Textile and Clothing Design Technology*, London and New York: CRC Press (part of Taylor and Francis group), 196–225.

Pond, G. (1973), *An Introduction to Lace*, New York: Charles Scribner's Sons.

Power, E. J. (2015), 'Yarn to Fabric: Knitting', in R. Sinclair (ed.), *Textiles and Fashion*, Cambridge: Woodhead, pp. 289–305.

Ray, S. C., and M. Blaga (2018), 'Warp-Knitted Fabrics', in T. Cassidy and P. Goswami (eds.), *Textile and Clothing Design Technology*, London and New York: CRC Press (part of Taylor and Francis group), pp. 196–225.

Spencer, D. J. (2001), *Knitting Technology*, Cambridge, UK: Woodhead, previously 1983, 1985, 1986, 1989, 1991 and 1993, with Pergamon Press.

Tubbs, M. C., and P. N. Daniels (eds.) (1991), *Textile Terms and Definitions*, ninth edition, Manchester: The Textile Institute, previously (with different editors) 1954 revised and enlarged 1955, 1957, 1960, 1963, 1967, 1968, 1970, 1972, 1975, 1978, 1984, 1986, 1988.

Walker, M. (1981), *Patchwork and Appliqué*, in the Good Housekeeping series, London: Ebury Press.

Walker, M. (1983), *Quilting and Patchwork*, in the Good Housekeeping series, London: Ebury Press.

6 Felt, bark and other nonwovens

6.1 INTRODUCTION

In the context of this present book, the collective term 'nonwovens' refers to all textile cloths not explained to some extent up to this point (so excludes cloths produced using weaving, knitting and the other techniques mentioned previously). By the early-twenty-first century, in terms of quantities used, nonwoven textiles created from natural, regenerated, synthetic or mixtures of fibres formed a surprisingly large proportion of all textiles consumed. Although there are various manufacturing techniques, one of two fundamental approaches may be taken: either fibres (in the form of a web, or batt, of regular thickness) are combined through mechanical entanglement, or fibres are bonded together using heat or the addition of some chemical or adhesive. Traditionally, the only nonwoven textile in widespread use was a felted cloth made from sheep's wool. Further types of nonwovens include needle-punched nonwovens; stitch-bonded nonwovens; wet-laid nonwovens; dry-laid nonwovens; air-laid nonwovens; adhesive-bonded nonwovens; spun-laid nonwovens; thermally bonded nonwovens; and hydro-entangled nonwovens. There appear to be several varieties of each, and by the early-twenty-first century it appeared that manufacturers (when assessed mainly through website-based information) were keen to identify types of nonwovens unique only to their product range; so there seemed to be the tendency to create unknown names for broadly familiar processes and products. In this present book, in addition to traditional wool felt, the focus is on explaining the types mentioned in this introduction, as it is believed that each important variety can be identified and explained under one of these general headings.

Although cloth-like products such as vellum, leather, fur, papyrus, feathers and paper are outside the subject context of the present book, nevertheless, in each case, a brief definition is given below in this introduction and key sources are identified also (as it could be argued that these product types bear a close relationship to felted or nonwoven textiles and that technological innovations, particularly in paper manufacture, inspired innovations also in nonwoven cloth manufacture).

Vellum (or parchment) is a smooth, durable writing surface, prepared from animal skin. Traditionally, the source was calfskin, though sheep and goat skins have been used also. Vellum often took the form of single pages or scrolls. The preparation involves cleaning, bleaching, stretching on a frame, scraping and finally treating the surface with lime or chalk to make it suited to writing or printing. A good source of further explanation was provided by Clements and Graham (2007).

Leather is a flexible and durable animal skin; hide or skin of cattle is the most commonly used. The skin needs first to be prepared for tanning and goes through several processes before it can be used further. A series of useful web-based links and various press releases are available at the website of an organisation known as Leathernaturally (http://www.leathernaturally.org/, accessed at 12 noon, Tuesday, 31 March 2020).

Fur includes both outer hair and skin of mammals, in years past used often as clothing; it may be regarded as a precursor to textile-based clothing. Further explanation was provided in the Encyclopaedia Britannica website (https://www.britanni ca.com/topic/fur-animal-skin#ref114471, accessed at 12:15 pm. Tuesday, 31 March 2020).

Papyrus is a writing material of plant origin, associated closely with ancient Egypt (dating from around the 4th millennium BCE). The source is from the papyrus plant, which was abundant across the Nile delta and seemingly throughout much of the Mediterranean region. Fuller explanation was given by Bell and Skeat (1935) and by Leach and Tait (2000).

Feathers are associated with birds and aid flight as well as waterproofing and insulation. Like wool fibres, they are predominately of protein. A substantial review was provided by Prum and Brush (2002).

Paper is a thin writing surface created from pulped cellulose fibres (invariably from wood). After pressing and drying, flexible sheets are formed that may be used as writing or printing surfaces, or as packaging material of various kinds. Attested by various archaeological finds, paper was invented in China during the second century CE. Further information was provided by Monro (2014).

It may be possible to consider further products alongside the non-textile products mentioned above, but it has been decided to ignore any further possibilities in order to ensure that the principal attention remains on fibre processing and the resultant textile products.

6.2   GENERAL TYPES

Nonwoven cloths are manufactured directly from staple fibres or continuous filaments and do not require the elaborate preparation associated typically with conventional woven or knitted textiles. The earliest form of nonwoven textile was wool felt, produced from layers of cleaned wool, with one layer often overlaid on another. When subjected to heat, moisture and agitation, the surface scales of the fibres tend to entangle. Occasionally a lightly felted surface is created on a fine-woven wool textile and this surface is raised (the baize cloth covering a typical gaming table is an example). Burkett's late-twentieth-century publication *The Art of the Felt Maker* is probably the most comprehensive attempt to identify past literature and to review different wool-felting techniques, both historically and geographically (Burkett, 1979).

Another type of textile produced by unconventional means is bark cloth. Once produced and used commonly in parts of Africa, Asia and across numerous islands in the Pacific, bark cloth was created from the inner bark of various trees, by beating the fibrous matter with a wooden mallet. Probably the best source for further information is a UNESCO Report, with a 10-minute video, entitled: 'Bark Cloth Making in Uganda' (UNESCO, 2008). A good comprehensive and readily understandable guide was provided also by Kooijman (1988).

It seems that all nonwoven cloths have the same starting point: a web or batt of fibres of similar density throughout. The term 'web' is often used to refer to a thin layer of fibres (like a sheet of fibres) and the term 'batt' to refer to circumstances where two or more webs are combined, though, throughout the literature, the two terms appear to be used interchangeably. Within this present book, where a layer of fibres has been produced and this is processed without further combination, the term 'web' is used, and where combination with other layers is a feature before further processing, the term 'batt' is used.

Under the heading 'nonwoven' in *Textile Terms and Definitions*, Tubbs and Daniels observed that 'Opinions vary as to the range of ... [cloths] to be classified as nonwoven. In general, they can be defined as textile structures made directly from fibre rather than yarn' (1991: 211). Adopting a definition for a nonwoven textile, sourced mainly from the American Society for Testing and Materials (under ASTM D117-80), Tubbs and Daniels gave the following alternative: 'A textile structure produced by bonding or interlocking of fibres, or both, accomplished by mechanical, chemical, thermal, or solvent means and combinations thereof' (Tubbs and Daniels, 1991: 212). This latter definition is acceptable as an explanation of a nonwoven textile used in this present book.

Webs of fibres can be dry-laid, wet-laid or produced by the spun-melt process prior to bonding (a term used to denote the strengthening of the web). Dry-laid nonwovens are of several types and, in each case, could involve two or more webs of fibre. Three avenues seem apparent. First, carded webs can be placed, one on top of another, to create what is known as a parallel-laid batt (indicating a parallel feature of the webs themselves rather than the constituent fibre orientation of each). Second, webs can be cross-laid, with one web placed at an angle to the web above and/or below. Third, fibres can be made into a web in order to remove any preferential fibre orientation resulting from the process. In this case, loose fibres are carried in currents of air and then deposited on a mesh conveyor to create the web in the form of a random-laid sheet (Wynne, 1997: 210–211). With this third type, known as air-laid nonwovens, opened fibres (including very short fibres not suited for conventional spinning) are mixed with an air stream and laid on a metal mesh in advance of some method of entanglement or bonding.

Water-laid or (wet-laid) nonwovens are produced by suspending fibres in a sort of slurry (like the manufacture of paper) and then depositing the fibres (which have taken up a random orientation) on a moving wire screen and, after draining, forming a web which is further consolidated by passing between rollers.

Various spun-melt systems are available also for webs of fibres in continuous-filament form (though an adhesive-bonding route mentioned in the next section may be used also in order to further consolidate constituent fibres). With continuous-filament webs, two systems are of relevance: spun-laid and melt-blown. In the former case, filaments, extruded through a series of spinnerets, are stretched and cooled before being deposited as a web (with randomly oriented filaments) on to a conveyor. This web can then be transported to a bonding stage. Meanwhile, with melt-blown a similar procedure to spun-laid bonding is followed initially, but filaments are stretched more intensively in a high-velocity air stream after extrusion from the spinnerets to enable much finer filaments to be produced in the deposited web.

6.3 TECHNIQUES AND VARIATIONS

There are several means of bonding fibres together, to produce nonwoven cloth types. Historically, in relation to wool this meant pressed felts, but in the modern nonwovens industry the most common variants are needle-punched nonwovens; stitch-bonded nonwovens; hydro-entangled nonwovens; adhesive-bonded nonwovens; and thermally bonded nonwovens. Further attention is focused on each below.

While the production of sheep's wool felt (in more modern times, often referred to as pressed felt) continued down through the centuries, alternative means of bringing collections of fibres together evolved. Initially, during the nineteenth century these developments were focused on wool and other animal fibres but, by the late-twentieth century, extended across all fibre types, both natural and manufactured. Mechanical entanglement, involving punching needles through a wad of wool (from one side or possibly both), was introduced as a cost-effective alternative to traditional felted wool in the 1840s (Thompson and Thompson, 2014: 152). This technique continued to be developed and, by the early-twenty-first century, was used with the full range of available fibres and blends. Needled or needle-punched cloths are produced (on a needle-punching machine) by pushing barbed needles through the full width of a web of fibres, thus forcing some of the constituent fibres to other parts of the web where they remain when the needles are withdrawn. Invariably, needles (with up to 2,000 on each machine) are triangular in cross-section with three barbs spaced along each of the three angles of each needle (Wynne, 1997: 214). Properties of the resultant cloth depend largely on needle penetration and needle density.

With stitch-bonded cloth, fibres are associated or bonded with each other through columns of stitches penetrating the fibre batt. Generally, stitch bonding uses a cross-laid batt, and often the machine used to apply the stitching is a modified warp-knitting machine (Wynne, 1997: 215).

Entanglement using jets of pressurised water is another possibility (a technique referred to as hydro-entanglement), though further association of constituent fibres was often encouraged through the application of an adhesive (outlined below) to ensure full bonding of the fibres within the web/batt (Wynne, 1997: 215). Hydro-entangled nonwovens are produced with pressurised jets of water, directed into webs or batts of fibre. Entanglement of fibres takes place when the jets of water penetrate.

An adhesive-bonded cloth consists of a web or batt of fibres held in close association (or bonded) by adhesive material (Tubbs and Daniels 1991: 2–3). With adhesive-bonded nonwoven cloths, dry-laid fibres are impregnated by an adhesive-latex binder; the type of adhesive and its concentration will influence the cloth's characteristics. A textile cloth rarely mentioned in general textile textbooks is laminated cloth. This is a cloth consisting of two or more cloth components, often bonded together through the addition of an adhesive.

Melt- or thermal-bonded nonwovens involve the fusion of thermoplastic fibres through the application of heat. Blends of compatible thermoplastic and non-thermoplastic fibres are possible also. As with all the other bonding techniques, several possibilities arise, including the use of fibres which bond together simply with the application of heat and pressure, or possibly with the use of a bonding agent (often in powder form) which reacts on application of heat. Also, the use of bi-component fibres is another possibility. These are man-made fibres with more than one constituent polymer made either side by side, or with one fibre forming the core of the other or with one fibre forming islands in the sea (in enlarged microscopic cross-section, typically, one fibre can be seen as a series of circles within the larger circle of the other fibre). For thermal bonding a bi-component fibre normally consists of two thermoplastic polymers with two different melting points. On heating, the low-melting-point polymer flows and then cools to bind adjacent fibres together, while the high-melting-point polymer remains intact as a structural support.

It can be seen, therefore, that an impressively wide range of manufacturing techniques is associated with the production of nonwoven cloths. Choices of constituent fibres can be made from the full spectrum available. It is not surprising that a vast range of desirable product characteristics can be addressed.

6.4   CHARACTERISTICS AND PROPERTIES

It appears that all fibres can be brought together in nonwoven form, but the choice of fibre and technique of nonwoven manufacture are dependent on the performance requirements of the final cloth, as well as anticipated costs and similar targets. Nonwoven products offer an amazingly wide range of properties. By the second decade of the twenty-first century, the period of emulating other conventional textiles was long gone, and nonwoven cloths could address numerous uses reserved previously for non-fibrous materials. Nonwovens could resist abrasion, rot, mildew, chemicals, flames, creases and bullets. They could be strong, tear-resistant, stable, soft, stiff and washable. They could be elastic and stretchable, ironable, dyeable, biodegradable, breathable, durable or throw away, foldable and absorbent. Yet these characteristics are not held within one fibre type and one technique of manufacture. Rather each characteristic needs to be selected depending on the anticipated end use and, based on this selection, suitable fibres and techniques must be selected also.

By the early-twenty-first century, nonwoven cloths were produced in large quantities, and a very wide range of product applications could be noted, including disposable applications such as female-hygiene products, bandages for wound dressing and wet wipes, and more durable applications such as insulation felts, wall coverings, carpet backing and crop covers. Further attention is focused on nonwoven product end uses below.

6.5   PRODUCTS

Nonwoven end uses are numerous. Elsasser, writing in 2005, using material given by the Association of the Nonwoven Fabrics Industry (2005: 155), listed several end uses under each of twelve general areas: agriculture; automotive; civil engineering; clothing; construction; interior furnishings; household; industrial and military; leisure and travel; health care; personal care and hygiene; and school and office. Elsasser's observation that nonwoven cloths were the fastest-growing segment of the textile industry appears to be the case also at the end of the second decade of the twenty-first century.

Up to the early-twenty-first century, agricultural end uses were mainly as covers which helped to retain heat in the ground and to avoid insect damage to crops. Nonwoven cloths were used also to hold strips of seed (so seeds were planted a predetermined distance and did not require distance estimates from the planter) as well as to provide weed control. In the automotive industry, nonwovens had taken up a multitude of roles, and resisted sound, heat and vibration, not to mention use as components of interior trim throughout, upholstery and as carpet backing, as well as oil and air filters. Many of the product avenues taken up in the automotive industry were of relevance also in the aeronautical industry, with uses throughout the interior as well as for engine filters of various kinds. Several uses in civil engineering and construction had become apparent, and involved numerous avenues of soil and landscape stabilisation, drainage, strengthening and erosion control as well as uses such as roofing underlay, pond linings and domestic sound insulation. Clothing functions had expanded substantially also and covered an amazingly wide spectrum of end uses, including applications extending from clothing interfaces to high-performance wear, underwear, paddings of various kinds, swimwear, as well as skiwear, rainwear and numerous avenues of personal protection. Uses in interior furnishings extended from wall coverings, carpet backing, pillows and pillowcases and components of upholstery to curtaining, quilts and blankets. Uses across household products were extensive, including numerous sorts of wipes, kitchen and wash cloths of various kinds, heat-resistant cloths, vacuum-cleaner dust bags, tea bags, coffee filters, laundry bags, ironing-board covers, tablecloths and napkins. In the industrial context nonwovens were used extensively as filters and insulation of various kinds as well as in areas associated with the general health and safety of the workforce, especially in protective clothing. In the military context, uses extended from protective clothing, bags of various kinds, to bullet-proofing and the production of parachutes. In leisure, travel and sports, uses included boat sails, tents, fibreglass boats, sleeping bags, headrests, blankets, footwear, rainwear and cloth covers of various kinds. Health-care uses were extensive and included surgical gowns, masks, caps and shoe covers, as well as bandages, bed linen and sterile packaging of various kinds. Personal-care and hygiene uses included children's nappies, disposable underwear and various incontinence products as well as wet and dry wipes. School and office use included book covers, envelopes, towels and various promotional items.

6.6 SUMMARY

This chapter was concerned with various forms of cloth known collectively as nonwovens, including wool felts and textiles made from the fibrous parts of tree bark, with mention made also of non-textile products such as vellum, leather, fur, papyrus, feathers and paper. It was explained that conventional wool-felt manufacture used agitation, heat and moisture to entangle the fibres and this was made possible through the scaled surface of the constituent fibres. Brief reviews were presented of relevant processing techniques and resultant products.

The anticipated end uses for nonwoven textile structures developed over the late-twentieth and early-twenty-first centuries were numerous. Twelve general areas of use can be identified: agriculture; automotive; civil engineering; clothing; construction; interior furnishings; household; industrial and military; leisure and travel; health care; personal care and hygiene; and school and office. At the beginning of the third decade of the twenty-first century, it appeared that nonwoven forms of cloth, in terms of weight of fibres processed, would soon (within a decade) be significantly more numerous than for any other cloth-producing sector globally.

REFERENCES

Bell, H. I., and T. C. Skeat (1935), *Papyrus and Its Uses*, London: British Museum pamphlets.

Burkett, M. E. (1979), *The Art of the Felt Maker*, Kendal, Cumbria, UK: Abbot Hall Art Gallery.

Clements, R., and T. Graham (2007), *Introduction to Manuscript Studies*, Ithaca: Cornell University Press.

Elsasser, V. H. (2005), *Textiles. Concepts and Principles*, second edition, New York: Fairchild.

Kooijman, S. (1988), *Polynesian Barkcloth*, Princes Risborough, Buckinghamshire, UK: Shire Publications.

Leach, B., and W. J. Tait (2000), 'Papyrus', in P. T. Nicholson and I. Shaw (eds.), *Ancient Egypt Materials and Technology*, Cambridge: Cambridge University Press.

Monro, A. (2014), *The Paper Trail: An Unexpected History of the World's Greatest Invention*, London: Allen Lane.

Prum, R. O., and A. H. Brush (2002), 'The Evolutionary Origin and Diversification of Feathers'. *The Quarterly Review of Biology*, 77(3):261–295.

Thompson, R., and M. Thompson (2014), *Manufacturing Processes for Textile and Fashion Design Professionals*, London: Thames and Hudson.

Tubbs, M. C., and P. N. Daniels (eds.) (1991), *Textile Terms and Definitions*, Manchester: The Textile Institute.

Wynne, A. (1997), *Textiles*, London: Macmillan Education.

WEBSITE REFERENCES

UNESCO website (2008), Bark Cloth Making in Uganda, Report plus ten-minute video (accessed 30 July 2019): http://www.unesco.org/archives/multimedia/?s=films_details&pg=33&id=641

Leather Naturally website (accessed 31 March 2020): http://www.leathernaturally.org/

Encyclopaedia Britannica website (accessed 31 March 2020): https://www.britannica.com/topic/fur-animal-skin#ref114471

7 Dyes and their application

7.1 INTRODUCTION

The colouration of textiles is through a process known as dyeing. Most textiles can be dyed, though techniques differ from one situation to another. Colour on textiles should be even (or level) and have a high degree of permanence (or should be fast) to its anticipated environment, and there are several approaches to achieving this. A wide variety of dyestuffs, techniques and machinery is available to manufacturers. Most textile dyeing is done either in dye houses, associated with large vertically organised textile producers, or by relatively small-scale specialist dyers, often working on a commission basis. In the early-twenty-first century, the vast bulk of the textile industry globally was using synthetic dyestuffs; practical knowledge of natural dyestuffs and their application was largely long gone from manufacturing memory. The intention of this chapter is to identify the principal dye classes and the types of machinery used, and to present an outline of the procedures necessary in the colouration of textiles, including various forms of printing. While the focus is on textiles and their dyeing and printing, the reader could benefit from consulting *The Printmaking Handbook* by Woods (2008), which gives a detailed breakdown of processes and products of value to the general artistic practitioner and designer, though the book is not focused specifically on textile colouration. Among the best publications with a textile focus are Foulds (1990), Storey (1992), Miles (1994), Ujiie (2006) and Wisbrun (2011).

7.2 DYE TYPES

Dyes are classed as either natural or synthetic. As their name implies, the former class are derived from plant, animal or mineral sources, though the clear majority of natural dyes are from plant sources. To ensure that a natural dye fixes to the cloth, a mordant is invariably added; this addition combines with the dye on the cloth to form an insoluble compound. Indigo, woad and walnut husks appear to be the only natural dyes which do not require the addition of a mordant to ensure fastness. Examples of further natural dyes include saffron, logwood chips and madder. Probably the most important step forward in textile dyeing over the past few thousand years was the discovery of the first synthetic dye by William Henry Perkin in the mid-nineteenth century. This first aniline dyestuff (which became known, in Britain at least, as 'Perkin's mauve' or, occasionally, 'Perkin's purple') heralded numerous subsequent developments. Crucial properties of dyestuffs include: the ability to display intense colour; solubility in an aqueous solution (at least during the period of dyeing); substantivity or the ability to be absorbed by the fibres being dyed; fastness or the ability to be retained by these fibres during the anticipated use (Giles, 1971: 27).

In the early-twenty-first century, dyes were sold in powder or granular form and occasionally as pastes or liquids. Often, textile dyeing was conducted in a water bath, though other methods relied on the use of other solvents. Certain dyes have an affinity for certain fibre types; if so, these dyes are substantive, that is, they are absorbed by the fibres but need further attention if they are to bind with the fibre and would not be released as easily as they were taken up.

There are numerous synthetic dyes. Probably the most important are acid dyes; azoic dyes; basic dyes; chrome or mordant dyes; direct dyes; disperse dyes; reactive dyes; sulphur dyes; vat dyes; pigments. The suitability of each to particular fibre types is identified below. Acid dyes can be used in the dyeing of wool, specialty hair fibres, silk and polyamide fibres. Azoic dyes are used mainly for cellulosic fibres. Basic dyes, first developed during the nineteenth century, have particularly pure and brilliant colours on fibres such as silk (and other animal fibres) but have very poor all-round fastness (Storey, 1978: 75) and, by the late-twentieth century, were only used to a limited extent in textile dyeing, especially of acrylics. Chrome dyes are used widely in wool dyeing. They are simple in application and retain good light and wet fastness. Direct dyes are regarded as the simplest class of dyes. They are readily soluble in water and are used mainly to dye natural cellulosic fibres as well as viscose. Fibrous material is simply placed in a weak solution and heated. Exhaustion of the direct dye improves with the presence of common salt, and an increase in temperature can increase the rate of dye uptake. Giles described various after-treatments which would help to improve the wet fastness of direct dyes (Giles, 1971: 61). Disperse dyes are suited to secondary acetate, triacetate and polyamide fibres, as well as acrylic and polyester fibres. Disperse dyes are regarded as the main dye class for synthetic fibres. Reactive dyes are used mainly for cellulosic and wool fibres but have been used with some success in the dyeing of silk also. Further details were provided by Giles (1971: 73). Sulphur dyes are used widely in a soluble form on cottons and provide fast colours. Vat dyes are used mainly for cellulosic fibres and have excellent all-round fastness. Pigments are suited to all fibres and their combinations. A summary of the main dye types and their properties was given in tabular form by Wynne (1997: 259).

The main dye groups used for specific fibre types are as follows: wool and specialty hair fibres are best dyed using acid or reactive dyes; silk is best dyed using acid, direct or reactive dyes; cotton, flax, jute and viscose are best dyed using direct, vat, azoic, sulphur or reactive dyes; acetate is best dyed using disperse dyes; nylon is best dyed using acid or disperse dyes; acrylics are best dyed using basic or disperse dyes; polyesters are best dyed using disperse dyes. Pigment dyes, suited to most fibre types, do not penetrate the fibres but rather bind with the surface. Although pigment colours may have reasonable wet fastness, they have bad rub fastness and the resultant textile is rather stiff. While dyes are often classed by reference to their applicable fibre type, the means of application and the machinery available are crucial also.

7.3 DYE TECHNIQUES

Prior to dyeing, cloths need to be prepared to ensure that impurities (possibly added during weaving or maybe surviving from the spinning stages) are removed as these may prevent dye uptake to certain areas. This process is known as scouring and involves soaking the cloth in a warm or boiling solution containing a commercial (PH neutral) detergent. By way of after-treatment, cloths are rinsed thoroughly in clean water and dried. Some cloths may also undergo a bleaching process prior to dye application, with an alkaline-type bleaching bath for cotton and linen and an acid-type bath for wool and silk. A series of scouring and bleaching recipes for various fibre types was given by Kinnersly-Taylor (2004: 34).

Dyeing can take place at one of several stages during textile processing: prior to the extrusion stage, at the loose fibres stage, at the sliver stage, at the yarn stage or at the cloth stage. Dyeing fibres prior to extrusion may also be referred to as polymer, spun or dope dyeing. With this process, a dye is added to the liquid to be extruded, so that the colour is firmly locked inside the fibre and is fully fast; if the fibre is not damaged the colour remains safe (Humphries, 2004: 211). 'Stock dyeing' is the term given to fibre dyeing. Top dyeing at the sliver stage, yarn dyeing at the yarn stage and piece dyeing at the cloth stage are the further alternatives. (It should be noted in passing that the word 'top' is used normally to denote a wool sliver of particularly high quality, with fibres highly aligned and parallel, achieved after a combing process.) Various dyeing machines are available, and selection is often based on the type and quantity of textile material being processed, and whether it is in loose fibre form, sliver form, yarn form or fabric form. Yarn dyeing takes place prior to weaving or knitting and is often used to create stripes and checks. Yarn dyeing may be in the form of packages (on a yarn holder of some kind which is preferable to skein or hank dyeing where yarns are looser), prior to warp or weft creation, or may occur to the warp yarns when they have been wound on to a warp beam. Dyeing at the stage of the warp beam offers greater economy compared to package dyeing. Stock dyeing and top dyeing are often used by tweed manufacturers, where the required colour effect is best achieved through mixtures of differently coloured fibres. Dyeing in the piece is the most common dyeing method, where cloth is simply prepared, dyed and finished. A further possibility is dyeing at garment stage, where a complete garment is dyed; this is more appropriate to simpler garments (such as T-shirts, sweatshirts and hosiery). Garment dyeing of tailored items such as dresses and garments appears to be avoided, probably due to considerations such as the effect on thread colour, zippers and other additions, as well as the possibility of different degrees of shrinkage to different parts of the garment. A summary of the advantages and disadvantages of dyeing at different stages of textile processing was given by Elsasser (2005: 177). In the early-twenty-first century, the vast bulk of cloth was dyed in the piece.

Dyeing may be completed in batches or may be continuous in nature. With the batch-dyeing method, a predetermined weight of cloth (maybe 100, 500 or 1,000 kilograms) is loaded into a dyeing machine containing dye and various dyeing auxiliaries (which assist dye penetration and fixation) dispersed in water, which is heated; when a certain temperature is reached, the dye molecules begin to migrate to the fibres (as they have a greater affinity to these rather than the dye solution). This migration may take a few minutes or a few hours (which varies depending on the type of dyestuff and fibres being used). Either the dye liquor is moved through the cloth or the cloth is moved through the dye liquor. A third possibility is to move both. A system known as pad dyeing is suited to both batch and continuous systems. In continuous dyeing, which (at the time of writing) accounted for well over half of the dyeing worldwide, the cloth is fed continuously into a dyeing machine. The intention with both systems is to ensure efficient dye fixation (a measure of the amount of dyestuff, which attaches itself to the textile, otherwise dye is wasted). Batch and continuous dyeing each require different amounts of dyestuff per unit weight of textile. Dye fixation is more rapid with continuous dyeing than with batch dyeing. Continuous dyeing is preferable when a large quantity of cloth needs to be dyed to the same colour. A major environmental consideration is the amount of dyestuff and auxiliaries required to dye the cloth to the required shade, and the amount of each that remains after dyeing takes place.

Four common dyeing-machine types include the jigger (or jig), the winch (or beck), the beam and the jet. Each is explained briefly here. There are two types of jigger, commonly used for cotton-cloth dyeing; one is opened and the other closed. The jigger operates with a low volume of liquor compared to the amount of fibrous material (a low material to liquor ratio of around 1:5 or 1:6 seems typical). The principal advantage is that this system permits cloth to be dyed at full width (so much creasing is avoided); chemical and heat losses are lower than with winch-dyeing machines. The major disadvantage is that a significant tension is exercised in the warp-ways direction; so, knitted cloths and delicate woollen and silk cloths are not suited for dyeing on the jigger. Winch (or beck) machines are the oldest dyeing machines; the cloth is stitched end to end and is circulated through the dye liquor. The machines are simply made, economical to purchase and relatively easy to operate. Further to this they are suited to most cloth types, especially those resistant to creasing when processed in rope form (i.e. after the cloth ends have been stitched together). The winch machine, when in use, will have less length-ways tension than the jigger machine and so is immediately preferable to use with various cloth types. It is also successful as a scouring machine. It can operate at maximum temperatures of around 95 degrees Celsius but will have a relatively high cloth-to-liquor ratio of around 1:20. Beam dyeing (replaced largely by the jet-dyeing machine, described towards the end of this paragraph) can be used to dye both yarn and cloth in open width. Cloth or yarn is rolled on to a perforated beam and is subsequently placed into the machine which is then closed and pressurised. The dye liquor circulates through the perforations in the beam. Under pressure, temperatures can be as high as 130 degrees Celsius. The machine was introduced initially in the mid-twentieth century to accommodate the dyeing requirements of relatively delicate warp-knitted nylon and acetate cloths. With the jet machine, high-temperature dyeing in rope form was introduced. The main advantages are that: dyeing time is much shorter when compared to other methods; dye penetration is good; low lengthways tension is present; the machine requires relatively small amounts of water; the material to liquor ratio, at around 1:5, is favourable;

production rates are high compared to other machines. The principal disadvantage is the high cost of purchase and maintenance.

A well-focused review of dyeing mechanisms and dyeing quality as well as aspects of dyehouse automation was provided by Shamey and Zhao (2014: 1–30); this source may be of interest to the less scientifically aware reader (though Shamey and Zhao's text does assume some previous knowledge of textile processing).

7.4   SCREENS AND OTHER MEANS

In the main, the range of dyes available for textile dyeing is available (after thickening) also to print textiles. Dyes and pigments are thickened by the addition of various gums and this provides further control over what is known as the print paste. Occasionally, after printing, a so-called wet print needs to be steamed and washed out in order to fix the colour and remove excess colour, respectively. With so-called dry prints (textiles printed using pigments), on the other hand, washing is not required. Printed embellishments may be applied to yarns, cloths and, occasionally, to full garments.

Various textile-printing methods have been developed over the years. Most common are block printing, engraved-plate printing, engraved-roller printing, flat-bed screen printing, rotary screen printing, transfer printing and digital printing. Attention is focused on digital forms of textile printing in section 7.5, but at this stage it should be stated that the development and introduction of digitally controlled printing techniques to textiles during the early-twenty-first century represented a far-reaching and exceedingly important technological innovation. Other textile-printing methods are explained briefly below in this section.

With block printing, probably the oldest textile-printing technique, a design is drawn on the surface of a flat wooden block and the area of the block around the design is carved away. The remaining raised area is covered with dye paste and then pressed against the cloth (which has been rolled out on to a flat area and is held under moderate tension). One carved block is required for each colour. The principal advantage of the method is the attractiveness of the result, which was in high market demand during the early-twenty-first century. The major disadvantage is the high skill level required of the workforce. A wide-ranging review of block printing is provided in the next chapter of the present book.

A system known as engraved-plate printing uses a flat engraved plate (probably of copper). This is known as an intaglio process where the design is 'incised or etched into the surface' (Woods, 2008: 20). This is unlike wood-block printing where the area left over after carving holds the colour to be printed. With engraved-plate printing, exceedingly detailed prints are possible, though often these are in a single colour (as registration difficulties are common with multiple colour prints, and it is difficult to ensure that second and subsequent colours are printed where intended on the cloth's surface).

An engraved-roller printing machine consists of a series of engraved rollers placed around a much larger metal cylinder. Each roller consists of a hollow steel cylinder electroplated with a coating of copper (the surface to be engraved) and is further chromium plated to prolong life. One roller is required for each colour to be printed. The engraved area of each roller is filled with a thickened dye paste. The cloth passes around the larger cylinder and picks up colour (and a part of the design) from each engraved roller. The technique can produce exceedingly fine-detailed effects, and because of its circular nature is capable also of printing continuous lines or similar effects in a warp-ways direction. The major disadvantage is the long machine downtime between designs, as well as the high expense of engraving a roller for each colour of a design. For economic reasons, this method was suited only to long print runs and, by the early-twenty-first century, had fallen out of use worldwide.

With flat-bed-screen printing, a fine meshed cloth (often of polyester, though in years past of silk, thus the name silk-screen printing) is stretched on a rectangular screen (of aluminium or wood) and coated with a photosensitive coating. Designs are rationalised into the minimum number of separate colours and a positive stencil is created on one screen for each colour. In each case, the stencil template is in the form of a sheet of clear acetate film, painted with an emulsion or ink through which light will not penetrate. Once painted, this acetate film is placed on the screen (which has been coated with a photosensitive coating) and is subjected to a high-intensity UV light source which hardens the photosensitive coating where exposed (unpainted areas of the clear acetate film). The remaining photosensitive coating remains soft and is washed out, thus leaving a permanent stencil on the screen. Each colour to be printed will require a separate screen. The first screen is laid on the cloth to be printed and dyestuff in paste form is squeezed through the openings in the screen using a rubber (or metal) blade known as a squeegee. Successive colours are printed by successive screens. Although flat-bed screen printing by hand is slow, it is relatively easy to set up at low expense, so may be regarded as economical for small specialist orders. In the late-twentieth and early-twenty-first centuries, flat-bed screens were often used to print designs on to T-shirts. Degrees of automation, however, were focused largely on printing long lengths of cloth (rather than individual T-shirts), and many such innovations were available during the late-twentieth century. By the early-twenty-first century, flat-bed screen printing in automated form was regarded as suited to printing cloth lengths but was not regarded as suited to printing ready-made garments. There are various design limitations; among these is the inability to print a continuous line or similar element down the length of the cloth (in a warp-ways direction); this is due principally to a difficulty in connecting a printed line to a printed line without gap or overlap. As the number of colours is increased, so also are costs, as separate screens are needed for each colour. The process, even when fully automated, is slow compared to rotary screen printing.

The faster automated method, known as rotary screen printing, uses screens in tubular form, with the dye being pressed through holes in the tube as the cloth comes into contact. In the early-twenty-first century, more textiles were printed by this method worldwide than by any other. The major advantage is the high speed of production, though one screen is required for each colour. The design restriction on printing a continuous line or similar element in a warp-ways direction (a typical difficulty with flat-bed screen printing) is eliminated due to the cylindrical design of the screen.

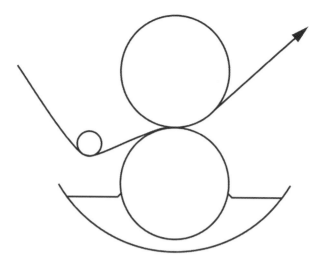

FIGURE 7.1   Cross-section view of simple padding mangle. Re-drawn by JW.

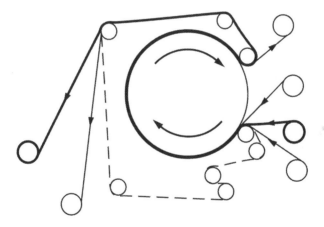

FIGURE 7.2   Continuous transfer printing. Re-drawn by JW.

With sublimation printing (often referred to as heat-transfer printing or thermal-transfer printing) disperse dyes are printed on paper which, on drying, is laid on the cloth and subjected to heat. At a high temperature, the dyestuff becomes vaporised and will have a greater affinity for the cloth than the paper and so will transfer to the cloth. The process is appropriate to all fibres with an affinity to disperse dyes, especially polyester, nylon, acrylic and acetate. The major advantage is that this is a dry process; the major disadvantage is the restriction to cloths predominantly of the fibre types mentioned. A major variant of sublimation printing is what is referred to as direct-to-cloth printing, a system which eliminates the advance printing of transfer paper, though pre-treated cloth and a post-print heating is required in order to ensure that the dye is made fast.

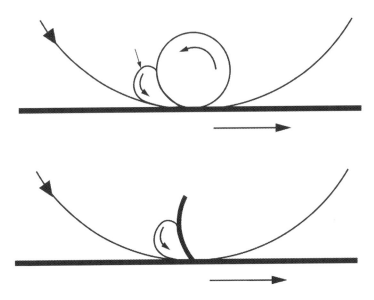

FIGURE 7.3   Rod or blade squeegees. Re-drawn by JW.

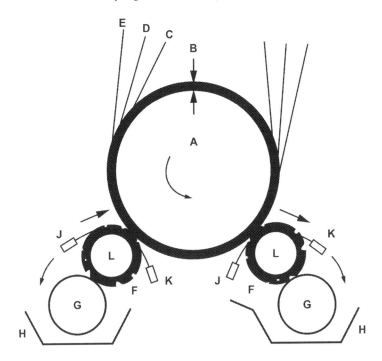

FIGURE 7.4   Two-colour engraved-roller printing (A = pressure bowl; B = resilient cover-ing; C = endless printing blanket; D = back-grey cloth; E = cloth being printed; F = engraved rollers; G = furnishing roller to transfer colour; H = colour box; J & K = colour doctors; L = inner steel mandrel). Re-drawn by JW.

FIGURE 7.5 Wooden block with steel inserts. Held at University of Leeds.

FIGURE 7.6 Roller-printed imitation, West Africa, mid-twentieth century, with label reading: 'guaranteed real Javanese deluxe'. Held at University of Leeds.

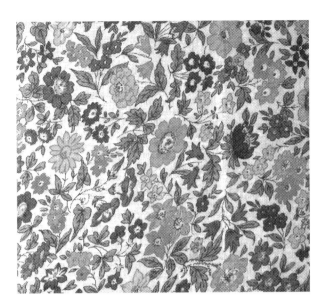

FIGURE 7.7   Early-twentieth-century roller-printed cotton, Carlisle (UK). Held at University of Leeds.

FIGURE 7.8   Mid-twentieth-century flat-bed screen-printed cotton. Held at University of Leeds.

Common terms used in textile printing include discharge printing, resist printing, duplex printing, warp printing, flock printing and devoré (or burn-out) printing. Discharge printing is a term used to refer to the removal (or discharge) of dye from areas of a previously piece-dyed cloth, often after printing these areas with a bleaching paste of some kind. Resist printing is when a resisting substance is applied to a cloth during printing in advance of placing the cloth in a dye bath. The dye will be

FIGURE 7.9   Early-twentieth-century roller-printed cotton (Shanghai). Held at University of Leeds.

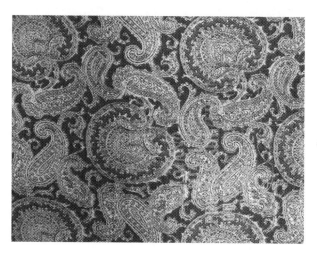

FIGURE 7.10   Early-twentieth-century roller-printed cotton (Carlisle, UK), in imitation of Paisley woven textile. Held at University of Leeds.

resisted in those areas printed previously with the resisting substance. Resist forms of colouring cloths are explained in chapter 8 of this present book. Duplex printing is when a design is printed on both sides of a cloth. Often the designs are the same and may appear to have resulted from weaving rather than printing. This method is very expensive and, by the early-twenty-first century, had fallen out of common use worldwide. With warp printing the warp yarns are printed (often using flat-bed screen printing) prior to weaving. This is exceedingly expensive, though costs are lowered if transfer printing of warps is used. The resultant cloth is similar to a warp-ikat cloth (described in the chapter 8 of this present book). With flock printing, an

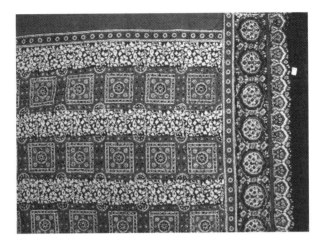

FIGURE 7.11 Late-twentieth-century screen-printed cotton, in imitation of block-printed cotton (Pakistan). Held at University of Leeds.

FIGURE 7.12 Late-twentieth-century screen-printed cotton, in imitation of block-printed cotton (Pakistan). Held at University of Leeds.

adhesive is printed on the cloth. Short fibres are then sprinkled on the cloth and stick to those areas printed with the adhesive. A more luxurious method, involving the application of an electrostatic charge (which can help to orient the short fibres vertically), is an exceedingly expensive alternative. A system known as devoré or burn-out printing is when areas of fibres within the cloth are printed with chemicals, which remove/destroy fibres within these areas. Invariably, the resultant cloth is weakened. In craft forms of textiles, further enhancement of a printed-cloth surface may be through hand painting or airbrushing.

FIGURE 7.13 Late-twentieth-century screen-printed cotton, in imitation of block-printed cotton (Pakistan). Held at University of Leeds.

FIGURE 7.14 Late-twentieth-century screen-printed cotton, in imitation of block-printed cotton (Pakistan). Held at University of Leeds.

7.5 DIGITAL PRINTING

Digital textile printing can certainly be regarded as a major innovation but, by the early-twenty-first century, the process still required major technical developments in order to lower total resultant costs substantially. Digital printing (referred to also as ink-jet printing) has several variants and the drop-on-demand (or DOD) technique is of relevance to textile printing.

FIGURE 7.15 Late-twentieth-century screen-printed cotton, in imitation of block-printed cotton (Pakistan). Held at University of Leeds.

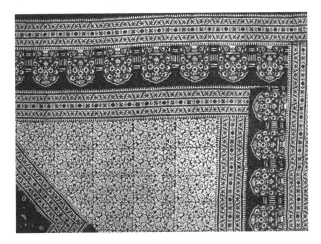

FIGURE 7.16 Late-twentieth-century screen-printed cotton, in imitation of block-printed cotton (Pakistan). Held at University of Leeds.

The major advantages of digital printing compared, for example, to rotary screen printing, include design flexibility (with the ability, for example, to produce large repeat sizes), suitability to short runs, the short time required to switch from one design to another, lower water usage, lower power consumption and lower space requirements. By the end of the second decade of the twenty-first century, the major disadvantage continued to be the relatively high installation price (compared to other textile-printing techniques) and the relatively low speed of output compared to rotary screen-printing methods.

Cahill, writing around the middle of the first decade of the twenty-first century, in his well-focused survey of early developments in digital printing (and associated

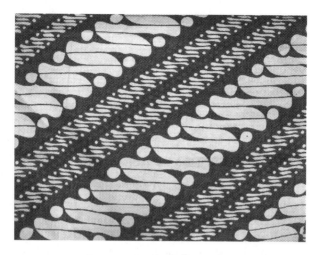

FIGURE 7.17   Late-twentieth-century screen-printed cotton, in imitation of Javanese batik (Java, Indonesia). Held at University of Leeds.

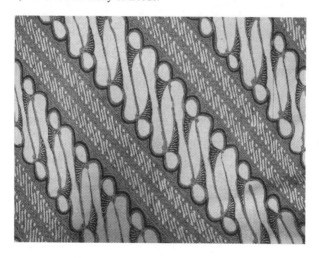

FIGURE 7.18   Late-twentieth-century screen-printed cotton, in imitation of Javanese batik (Java, Indonesia). Held at University of Leeds.

ink-jet techniques), concluded that 'Inkjet textile printing is growing while growth in analogue textile printing remains stagnant' (Cahill, 2006: 15). By the beginning of the third decade of the twenty-first century, this continued to be the case. Although most of the commentary appeared to recognise the benefits of digital forms of printing, the bulk of global textile printing was by means of rotary screen printing. By the beginning of the third decade of the twenty-first century, Italian textile printers were the principal European consumers of digital-printing machinery, and there was much scope for growth elsewhere, particularly in the Asia–Pacific region. The principal advantage was the immense design flexibility offered by the technique, and

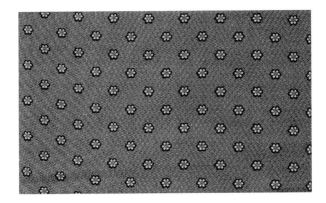

FIGURE 7.19   Mid-twentieth-century roller-printed cotton. Held at University of Leeds.

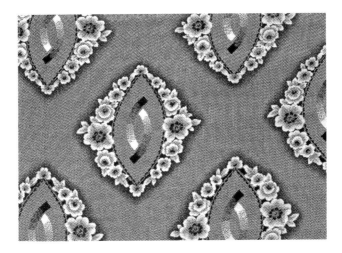

FIGURE 7.20   Mid-twentieth-century roller-printed cotton. Held at University of Leeds.

this continued to be the basis of market promotions in the early-twenty-first century. However, the major disadvantage continued to be high installation costs.

7.6   SUMMARY

Dyeing is a means of colouring textiles, invariably in a container known as a dye bath (which holds colouring matter). By the early-twenty-first century, at least in the mass-production context, natural dyes such as madder, indigo and saffron had fallen out of use and synthetic dyes had been substituted for these.

Printing, often used for further embellishment of textiles, is the application of dyes in paste form to localised areas of a cloth. Printing techniques include: block printing, engraved-plate printing, engraved-roller printing, flat-bed screen printing, rotary screen printing, transfer printing and digital printing. Most printed textiles are printed on one side of the cloth only, so colours on this face side are darker after printing. Probably the most far-reaching innovation associated with textile printing during the early-twenty-first century was digital printing; this offered numerous advantages compared to other methods. Probably the most significant of these were: the immediacy of results; the small scale of operations possible; lack of limitation on the number of colours; absence of an apparent limit on the size of the repeating unit. The further possibility of creating a non-repeating embellishment of some kind, possibly within a fashion context, was a further potential advantage (not listed, it seems, in the bulk of treatises dealing with the subject).

REFERENCES

Cahill, V. (2006), 'The Evolution and Progression of Digital Printing of Textiles', in H. Ujiie (ed.), *Digital Printing of Textiles*, Cambridge: Woodhead Publishing, pp. 1–15.

Elsasser, V. H. (2005), *Textiles. Concepts and Principles*, second edition, New York: Fairchild.

Foulds, J. (1990), *Dyeing and Printing*, small-scale textiles series, Rugby, UK: Intermediate Technology Publications.

Giles, C. H. (1971), *A Laboratory Course in Dyeing*, second edition (revised and enlarged), Bradford, UK: The Society of Dyers and Colourists.

Humphries, M. (2004), *Fabric Reference*, third edition, Upper Saddle River, NJ: Pearson and Prentice Hall.

Kinnersley-Taylor, J. (2004), *Dyeing and Screen-Printing on Textiles*, London: A & C Black.

Miles, L. W. C. (ed.) (1994), *Textile Printing*, second edition, Bradford, UK: The Society of Dyers and Colourists.

Shamey, R., and X. Zhao (2014), *Modelling, Simulation and Control of the Dyeing Process*, Cambridge, UK: Woodhead.

Storey, J. (1978), *Dyes and Fabrics*, London: Thames and Hudson.

Storey, J. (1992), *Textile Printing*, revised edition, London: Thames and Hudson.

Ujiie, H. (ed.) (2006), *Digital Printing of Textiles*, Cambridge: Woodhead Publishing.

Wisbrun, L. (2011), *The Complete Guide to Designing and Printing*, London: Bloomsbury.

Woods, L. (2008), *The Printmaking Handbook*, Edison, NJ, USA: Chartwell Books.

Wynne, A. (1997), *Textiles*, London: Macmillan Educational.

8 Block-printing and resist-dyeing techniques

8.1 INTRODUCTION

The technique of block printing using hand-carved wooden blocks encompasses various categories. In the main, this has involved one or a combination of the following processes: application of a discharge paste, which removes dye from areas of a previously piece-dyed cloth; direct application of a dye, invariably as a viscose paste; application of a mordant (again, as a paste), which allows dye to be accepted (only in the areas covered by the mordant); application of a resist paste of some kind, which prevents dye from reaching the areas covered. Largely, these processes rely on applications using hand-carved wooden blocks, though cut-paper and other types of stencil can serve a similar function. The focus throughout this chapter is on the embellishment of cotton cloths, though some of the processes are practised also on cloths of other fibres.

It should be noted at this stage that a particular topic dealt with under one heading here might have been examined under a different heading elsewhere. Ikat, regarded by some as a tie-and-dye technique, is dealt with as a separate section rather than being included under section 8.3 (entitled 'Tie-and-dye'). Common sense, tempered with the consensus in relevant publications to date, has been the guiding force for inclusion of the ikat technique within its own section.

Block printing was used in many geographical locations but had, by the late-twentieth century, been replaced largely by screen printing, though several craft producers in various locations were still using traditional ways and had not become part of the rush towards mechanised industrialisation. Some of these producers were still operational by the beginning of the third decade of the twenty-first century. Other forms of embellishment rely on covering or shielding parts of a yarn or cloth and immersing it in a dye bath (thus allowing the dye to take to those areas not covered). These techniques are known collectively as resist-dyeing techniques. The Ajanta cave paintings in western India are often given as evidence of very early use (and thus manufacture) of resist-dyed textiles, as it has been argued that various pieces of clothing on figures depicted in the caves were patterned using some form of resist technique. Probably the most convincing illustration from the Ajanta caves, depicting design types which were commonly obtained in the nineteenth and twentieth centuries, and which used ikat and tie-and-dye processes, was reproduced and presented by Karolia (2019: 212). It should be mentioned that, while an image on a cave wall is not absolute proof that the technique was in use and that the relevant cloths were being produced, it does, however, suggest familiarity among artists of the types of embellishments on cloths available at the time. The earliest part of the cave complex has been dated to the second century BCE, with further additions up to the late-fifth century CE (Madan, 1990: 173). So, there is indirect evidence for the use of resist-dyeing techniques sometime before the late-fifth century CE. There is also substantial literary evidence from the early centuries CE; much of this was reviewed briefly by Karolia (2019). Direct material evidence for resist and mordant dyeing in India from a much later date is substantial, mainly from the eighteenth century (CE) onwards, and in the form of large quantities of hand-block-printed and resist-dyed cloths in museums worldwide. Therefore, while the early development and use of resist- and mordant-dyed textiles in India seems likely, the immediate direct evidence within India itself is weak. Instead, early evidence in the form of excavated resist- and mordant-dyed textiles from archaeological sites in Egypt confirms an extensive trade across the Indian Ocean at least three hundred years prior to the arrival of European traders in India. Relevant studies on the Indian textiles found in Egypt were completed by Vogelsang-Eastwood (1990) and Barnes (1992 and 1997), and a more recent article was provided by Burke and Whitcomb (2004). After the arrival of Portuguese, Dutch and British traders, all in search of the rare spices of Asia, textiles acted as important exchange goods in many parts of Asia. There was, however, ample trade in textiles prior to the arrival of these European traders. Guy (1998) provided a well-illustrated review of the development of textiles in trade in Asia. Hann and Thomson (1993), Hann (2005) and Hann (ed.) (2008) provided brief reviews, in the Indonesian context, covering typical pattern types as well as processes.

Probably the most comprehensive review to date of painted, printed and resist-dyed textiles, including explanations of relevant methods and regional variations, was provided by Karolia (2019); although focused principally on cloths manufactured in India, ample useful explanation and illustration are given of all the relevant techniques. It should be noted that by the early-twenty-first century, producers in India, where numerous block-printing procedures had evolved over the years, offered the most extensive variety of embellished cloths worldwide. Karolia (2019: 34) identified the principal regions in India for textile printing and painting in the early-twenty-first century as: Gujarat, Rajasthan, Andhra Pradesh, Uttar Pradesh, Maharashtra, West Bengal, Odisha and Tamil Nadu. Interestingly, according to Karolia (2019: 34), although there had been much fusion of motifs, patterns and colours by the end of the second decade of the twenty-first century, expertly crafted products in many regions were still identifiable to that region.

The objectives of this chapter are to identify and explain the various applications associated with block printing, and to outline the nature of various resist forms of embellishment used on textiles. The focus is on handcraft techniques. Relevant literature is identified throughout.

8.2 BLOCK PRINTING

Block printing (using hand-carved wooden blocks) is probably the most ancient method of applying dyes (generally in paste form) to designated regions of a textile surface. Carved wooden blocks are often of teak and are of a size to allow easy manipulation. Dimensions of around 30 centimetres in length by around 25 centimetres in width and a depth of around 10 centimetres would be typical of a block described as large; often blocks are of around one third of this size. The process of producing a carved wooden block involves not just carving but also drilling and may take up to ten days to complete (Sreenivasam, 1989: 30). Various procedures in the production of carved wooden blocks were explained by Karolia (2019: 124–137); although methods and tools used vary from region to region, important considerations include: the type of wood selected; the preparation of the wood; transferring the design; carving; and finishing.

The block was invariably finished by adding a wooden handle to its top. Often the shape of the handle was an indication of who carved a block (Karolia, 2019: 133). Prior to being used for printing, newly carved blocks were soaked in oil as it was believed that this would prevent water absorption during printing (Sreenivasam, 1989: 30). Pre-treatment of a cloth prior to printing is crucial to ensure ultimate intensity and fastness of colours as well as even application. Before printing, scouring and bleaching, using local soaps, alkali and an animal dung, are typical and these make the cloth absorbent and aid the removal of impurities. A further pre-treatment prior to the application of a mordant may involve soaking the bleached cloth for a few minutes in a fusion of myrobalan fruit, water and castor oil; it was claimed that this would aid clear definition when printing.

Often the cloth is laid out under tension on a flat surface prior to printing. As indicated in the introduction to this chapter, four categories of application using carved wooden blocks can be identified: direct printing, resist printing, discharge printing, mordant printing. In her treatise concerned with cloth printing and dyeing, Wells gave numerous recipes suggesting approaches to discharge printing (Wells, 1997: 138–149). Most importantly, she highlighted exact times, temperatures and concentrations as well as potential difficulties when aiming for specific aesthetic effects (Wells, 1997).

In block printing, each colour is applied separately across the whole cloth, with successive colours only applied after the drying of the previous colour. Where more than one block is printed, it is often the case that pin marks are printed at each corner of the block and these are used as position markers to align the pins of each successive block and thus ensure effective registration from block to block and colour to colour. Fine details are often not printed from engraved wood, as wood would wear down or break easily. Rather, strips of metal (sometimes copper or brass) are wedged into the surface of the wooden block. In addition, often where wide expanses of a single colour are required, these areas on the block are covered with wool felt, which helps to ensure even coverage of the dye. Often air vents are drilled in the carved wood to allow the release of trapped bubbles which might otherwise gather in the block and lead to smudges in printing.

In the European context, it seems that the use of wooden blocks was first adopted in medieval times (around 1300). Centuries later, Morris and Company (under the influence of William Morris, the eminent British designer) adopted block-printing techniques and helped also in developing recipes, times and concentrations associated with the application of various natural dyes. This, in the wake of the Industrial

Revolution, was against a background where hand-carved block printing of textiles was being replaced across much of Europe by mechanised forms especially engraved-copper roller printing. Schaefer (1939) reviewed various early forms of printed textiles.

The purpose of this present section is to identify and discuss briefly the nature of three types of hand-block-printed cloths. Attention is focused on three product case studies (ajraks, kalamkari and dabu cloths), each regarded as traditional, with application by block using one, or combinations, of resist, mordant or discharge methods and, in one case (kalamkari), supplemented further with hand-painting techniques using a kalam. A kalam is a hand-held drawing implement about the size of a large fountain pen, with a handle of bamboo or similar material and a tip sometimes of pointed metal and sometimes of sharpened bamboo and including a wad of fibre or yarn acting as a reservoir between handle and tip. Karolia (2019: 97) and Guy (1998: 22) illustrated examples of relevant implements.

Ajrak cloths are traditional block-printed textiles, associated with the province of Sindh in southern Pakistan and Kutch (Gujarat, India); traditionally, the production of these cloths involved block printing of a resist material prior to dyeing. Ajraks often show a strict repetition, reminiscent of Islamic tiling design; so they consist largely of regular geometric patterns. In a well-illustrated treatise, Bilgrami (1990) provided explanations of production procedures as well as biographical details of numerous ajrak practitioners based in Sindh (Pakistan). By the late-twentieth century, in a few traditional hand-block-printing regions in both Sindh (Pakistan) and Gujarat (India), most ajrak production was aimed at addressing the perceived requirements of international buyers rather than for local consumption. Often, designs were reproduced not by traditional block-printing methods but rather by flat-bed screen in imitation of original traditional designs. Often these were crude aesthetically but were sufficiently like the originals to fool both international buyers and ultimate consumers. The brief outline given below relates especially to traditional procedures, which, although displaced largely in many regions by the early part of the third decade of the twenty-first century, still clung on in a few remote locations. As noted above, cotton ajrak cloths were produced traditionally in Sindh, as well as in a few locations in India including Kutch (in Gujarat); it is claimed that groups of families from Sindh settled in Gujarat in the sixteenth century and brought with them relevant textile-printing knowledge and skills. After a severe earthquake in 2001, many of the ajrak craft families from around the Kutch region moved eastwards to a village which became known as Ajrakhpur. Often ancient origins for ajrak are argued, with reference to the shawl depicted on a terracotta figure (known as the priest/king) associated with the ancient Indus Valley civilisation (dated variously, but 2,500–1,500 BCE is the most common time frame cited in the relevant literature), in the belief that the cloth depicted was of a type produced using an ajrak technique and also that the dominant motifs (each similar to a three-leaf clover) on the cloth were of a type associated with ajrak printing in more modern times (though the author has failed to find similar motifs in around 100 ajrak cloths examined to date). So, the evidence typically cited to support the argument for ancient origins of the ajrak technique is not convincing.

Dominant colours on ajrak cloths are indigo blue, traditionally from locally grown indigo; a rich crimson traditionally from madder; black outlines from various iron sources; white, provided by the bleached base cloth. Evidence for the ancient use of indigo (a key dyestuff used in ajrak production) in South Asia is substantial, and this is supported by the discovery of a dyer's workshop in the ruins of the city of Mohenjo-daro (a major ancient urban centre, now in Pakistan) dated to around 2,500 BCE. Various mid-twentieth-century publications provided worthwhile discussions of the extraction and use of indigo, including Vetterli (1951), Haller (1951) and Bühler (1951). Prior to printing, ajrak cloth went through several preparatory stages and, traditionally, a group of about ten cloths may have taken up to forty days to complete. End uses ranged from turbans and mantles of various kinds, to tablecloths and bed sheets. Traditionally, ajrak cloths were printed on both sides so would involve two sets of blocks, one set with the reflected images of the other. By the early-twenty-first century, vegetable dyestuffs, used traditionally, had been displaced by various synthetic equivalents. A good outline and discussion of ajrak preparation, production and use, including helpful illustrations, was given by Karolia (2019: 124–137).

In many international markets, kalamkari (or, occasionally qalamkari) cotton textiles became known as 'chintz', though, by the twentieth century, the term (chintz) was reserved for regular floral patterns on cotton, often glazed through a calendering process. Cloths were bleached first, washed, dried and sometimes pounded to make them ready for dye application. In ancient times, dyes were applied by hand using a kalam (or pen), or by wooden block or (most commonly, in the early-twenty-first century) using a mixture of both. When a cloth is described as a kalamkari, this indicates that a portion of the cloth was produced using a hand-painting technique. In each case, extracted colouring matter was ground into a fine powder, mixed with a binder or adhesive and applied to a textile surface. Traditionally, these textiles were used as scrolls, or temple hangings, or religious banners of various kinds depicting deities or scenes from Hindu epics such as the Ramayana or Mahabharata.

A region in Rajasthan famed for block-printed textiles is Bagru, a small village around 30 kilometres from Jaipur. Traditionally, dominant colours were primary red, maroon and black on a cream ground. Karolia observed that 'No reliable literature is available to indicate the beginning of this kind of printing, but it is said to have started 400 years ago' (Karolia, 2019: 184). Probably a key reason why textile printing developed in the region in the past was the plentiful supply of local water, suited to both dyeing and printing. Often river mud was a component of the resist material used prior to dyeing. Bagru is famed for a resist printing known as dabu, mud-resist hand-block-printed cottons, produced using black clay, acacia gum, spoiled wheat flour and limestone combined as a resist paste. Wooden blocks were carved from locally available wood with designs drawn initially on paper. When ready, these blocks were used to apply the resist paste; saw-dust was sprinkled over the resist after application. Once dried in the sun, the cloth was immersed in a cauldron of cold dye, and the dye would take to those areas not covered with the paste resist. After further drying in the sun, the cloth was washed thoroughly to remove excess dye as well as the mud-based resist. The non-dyed/resisted background was thus revealed. One dyeing process, using indigo, was common. Dabu cloths may show a veining effect;

FIGURE 8.1 Late-twentieth-century block-printed cotton (Pakistan). Held at University of Leeds.

as is the case with some wax-resist batiks, the resist may crack in the dyebath, allow the penetration of some dye and thus show unexpected irregular lines of colouring in the final cloth. Traditionally, dabu textiles were used as lehenga (a form of full-length skirt) and odhni (cloth used to cover the head and chest).

8.3 TIE-AND-DYE

Tie-and-dye (often referred to simply as tie-dye) is where a dye reaches only selected areas of a cloth's surface, and is prevented from penetrating other areas, because of the pressure maintained through tying or sewing. Variations are apparent worldwide, though it is Indonesia, India and Pakistan, as well as West Africa and Japan, whence knowledge of local variants has spread internationally, probably due to past trade and/or colonial connections. In the eighteenth and nineteenth centuries, printed and tie-and-dye textiles from parts of Asia, especially India, stimulated technological innovations in textile colouration among European and North American dye chemists as well as machinery manufacturers and, in more recent years, craft producers worldwide have adopted the relevant crafts themselves.

Rolling of a cloth (e.g. in a weft-ways direction) and then binding the roll tightly at regular intervals with a dye-resistant material (such as raffia, banana leaf or polypropylene string) will suffice to create a single-colour stripe design after immersion in a single dye bath, with the width of stripes dependent on the width of bound and

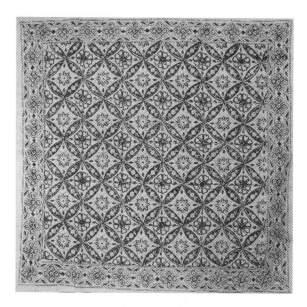

FIGURE 8.2 Late-twentieth-century block-printed cotton (Pakistan). Held at University of Leeds.

FIGURE 8.3 Late-twentieth-century block-printed cotton (Pakistan). Held at University of Leeds.

unbound areas. In Rajasthan and Gujarat (both in India) cloths produced like this are known as laheriya (meaning waves). After drying, a check-type design can be created by rolling the cloth and binding it in the opposite direction. Zigzag stripes are another possibility. Murphy and Crill (1991) provided a worthwhile review of types produced across much of India.

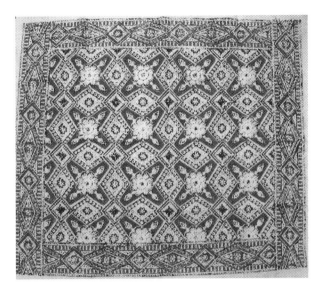

FIGURE 8.4   Late-twentieth-century block-printed cotton (Pakistan). Held at University of Leeds.

FIGURE 8.5   Late-twentieth-century block-printed cotton (Pakistan). Held at University of Leeds.

Numerous types of bound resist are possible; sometimes portions of a fabric are lifted and tied in such a way that circular-type forms are created after dyeing. Bühler observed that on occasions, small pebbles, ceramic or glass beads, grains of rice or seeds were tied in to give exceedingly fine results (Bühler, 1954). Commonly, in Indonesia such cloths are referred to as plangi, and, in Gujarat (India) as band-hani. The determination of the exact penetration of the dye is not possible. So, each item of tie-and-dye, like other craft products, is unique, a characteristic welcomed

FIGURE 8.6   Late-twentieth-century block-printed cotton (Pakistan). Held at University of Leeds.

FIGURE 8.7   Late-twentieth-century block-printed cotton (Pakistan). Held at the University of Leeds.

among Western consumers in the early-twenty-first century, yet frowned upon by many retail buyers (focused on mass-market needs) for much of the second half of the twentieth century. Bühler provided a relevant scholarly article dealing mainly with the variations in the technique and its distribution (Bühler, 1954).

Tie-and-dye variations are widespread throughout South Asia (mainly present-day India and Pakistan). Historically important areas were identified in the late-twentieth century by Westfall and Desai (1987). The simplest South Asian types show diamond- or circle-shaped areas of a few centimetres in width, repeating with

FIGURE 8.8   Late-twentieth-century block-printed cotton (Pakistan). Held at University of Leeds.

FIGURE 8.9   Screen-printed silk and cotton. Held at University of Leeds.

regularity over a dyed ground. Various methods can be used to guide the craftsperson tying the ties, including the use of stencils and blocks (Westfall and Desai, 1987; Else, 1988). Traditionally, products from each region had typical characteristics, designs and colours. For example, Bühler (1954), the renowned scholar on resist-dyeing techniques, observed that typical tie-dyed cloths produced in Rajasthan were embellished by wavy lines, zig-zags, crosses, leaves and various other simple motifs; largely, this is still the case at the time of writing. Also, more complex designs have

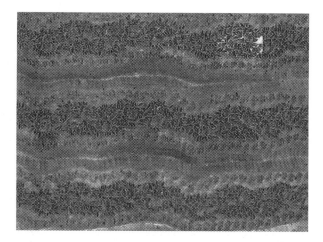

FIGURE 8.10   Wooden printing block, with metal additions. Held at University of Leeds.

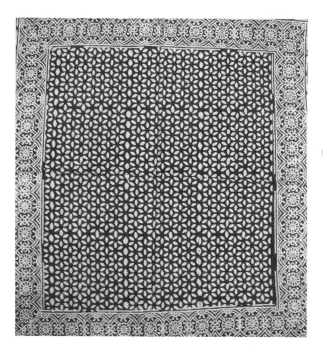

FIGURE 8.11   Block-printed discharge cotton print. Held at University of Leeds.

been produced in Rajasthan, and these feature more detailed outlines, various floral and plant forms as well as animal and human motifs (Bühler, 1954). End uses in India and Pakistan traditionally included saris, turbans, stoles and shawls. Often resist materials were left tied to the cloth prior to its sale as this indicated to the purchaser that this was indeed the genuine article rather than an imitation.

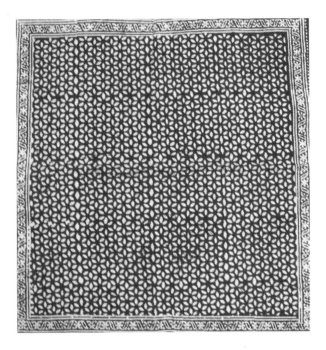

FIGURE 8.12   Block-printed discharge cotton print.

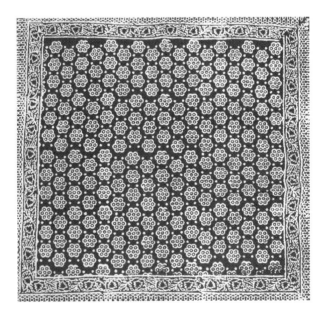

FIGURE 8.13   Block-printed discharge cotton print.

FIGURE 8.14 Block-printed discharge cotton print.

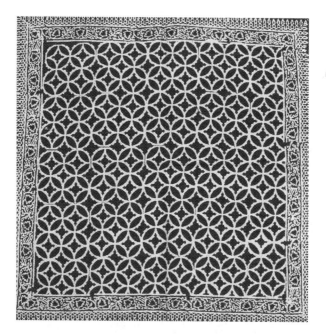

FIGURE 8.15 Block-printed discharge cotton print.

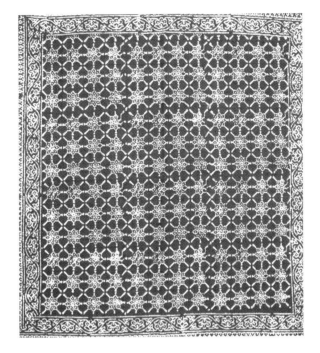

FIGURE 8.16   Block-printed discharge cotton print.

FIGURE 8.17   Block-printed resist on cotton. Held at University of Leeds.

FIGURE 8.18 Block-printed resist on cotton. Held at University of Leeds.

FIGURE 8.19 Block-printed cotton ajrak outlines. Drawn by JW.

FIGURE 8.20   Block-printed cotton ajrak outlines. Drawn by JW.

FIGURE 8.21   Block-printed cotton ajrak outlines. Drawn by JW.

FIGURE 8.22 Block-printed cotton ajrak outlines. Drawn by JW.

FIGURE 8.23 Block-printed cotton ajrak outlines. Drawn by JW.

FIGURE 8.24 Block-printed cotton ajrak outlines. Drawn by JW.

FIGURE 8.25 Block-printed cotton ajrak outlines. Drawn by JW.

FIGURE 8.26 Block-printed cotton ajrak outlines. Drawn by JW.

In Indonesia the most important regions for tie-and-dye cloths have traditionally been Sumatra, particularly around Palembang, and parts of Java, Lombok and Bali. Common motifs included circles, rings and lozenge shapes. Traditional end uses included shawls, sarongs, sashes, and wall hangings. An important tie-and-dye cloth variation typical of Indonesia is tritik which uses strong thread stitched in short lengths into the cloth and pulled tight to give closely packed folds, parts of which will resist dye when immersed in a dye bath.

The Japanese term 'shibori' is a catch-all term used to cover various tie-and-dye methods involving stitching, binding, pleating, gathering, folding or clamping of cloth, immersing it in a dye bath and thus allowing dye to penetrate only to certain parts of the cloth. After drying, with the removal of clamps or other additions made prior to dyeing, the dyed cloth is unfolded and laid flat. Traditionally, the technique was combined often with other resist-dyeing techniques as well as embroidery. Clamped resist, known as itajime, where the cloth is folded and clamped between two sticks or boards, is a rare form of tie-and-dye apparently unique to Japan. Southan (2008) provided an easily understandable and well-illustrated work of value to designers and visual artists wishing to begin experimentation with shibori-type imagery. Earlier reviews of relevant tie-and-dye applications in Japan were provided in the CIBA Review (Anon, 1967/4).

Various forms of tie-and-dye cloths were produced in West Africa, most notably among the Yoruba of Nigeria. Known collectively as adire cloth, these bound resists often featured large white circular elements against an indigo-dyed background; the large white circles were obtained by tying raffia around clumps of cloth prior to

FIGURE 8.27 Block-printed cotton ajrak outlines. Drawn by JW.

immersion in a dye bath. Further details were given by Picton and Mack (1989: 148). There is much evidence for traditional resist-dyed cloths in other parts of West Africa, including Senegal, Gambia and Sierra Leone. Tie-and-dye cloths were also produced in Iran, Syria and Cyprus. A good review of traditional geographical distribution was given by Bühler (1954). In pre-Columbian America, direct evidence is available to indicate production in Mexico and Peru. Traditional tie-and-dye cloths were also produced in parts of Cambodia, Myanmar and Thailand (Bühler, 1954). In the People's Republic of China, tie-and-dye techniques were associated traditionally with rural people in the south west, especially in Szechwan (sometimes spelled Sichuan) and Yunnan provinces, where the technique was associated with intricate folding and stitching in the production of blue designs on a white cotton background.

FIGURE 8.28 Block-printed cotton ajrak outlines. Drawn by JW.

FIGURE 8.29 Block-printed cotton ajrak outlines. Drawn by JW.

FIGURE 8.30   Block-printed cotton ajrak outlines. Drawn by JW.

FIGURE 8.31   Block-printed cotton ajrak outlines. Drawn by JW.

FIGURE 8.32 Block-printed cotton ajrak outlines. Drawn by JW.

FIGURE 8.33 Block-printed cotton ajrak outlines. Drawn by JW.

FIGURE 8.34   Block-printed cotton ajrak outlines. Drawn by JW.

FIGURE 8.35   Block-printed cotton ajrak outlines. Drawn by JW.

FIGURE 8.36   Block-printed cotton ajrak outlines. Drawn by JW.

FIGURE 8.37   Block-printed cotton ajrak outlines. Drawn by JW.

FIGURE 8.38   Block-printed cotton ajrak outlines. Drawn by JW.

FIGURE 8.39   Block-printed cotton ajrak outlines. Drawn by JW.

FIGURE 8.40   Block-printed cotton ajrak outlines. Drawn by JW.

FIGURE 8.41   Block-printed cotton ajrak outlines. Drawn by JW.

FIGURE 8.42   Block-printed cotton ajrak outlines. Drawn by JW.

FIGURE 8.43   Block-printed cotton ajrak outlines. Drawn by JW.

FIGURE 8.44   Detail of block-printed cotton ajrak, Pakistan. Held at University of Leeds.

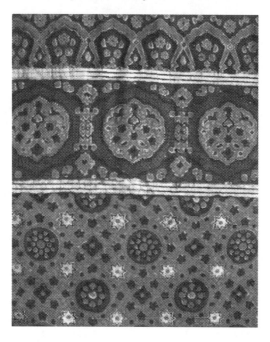

FIGURE 8.45   Detail of block-printed cotton ajrak, Pakistan. Held at University of Leeds.

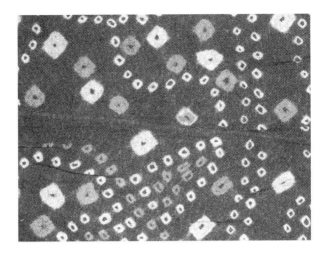

FIGURE 8.46 Detail of tie-and-dye cotton, mid-twentieth century, Pakistan. Held at University of Leeds.

FIGURE 8.47 Tie-and-dye cotton garment, mid-twentieth century, Pakistan. Held at University of Leeds.

Motifs included fish, flowers, birds and butterflies; end uses included children's wear, curtaining and bedcovers. Hann (2005) presented a brief review of resist-dyed cloth types and techniques.

8.4   WAX-, RICE- AND PASTE-RESIST TECHNIQUES

A typical resist material is wax, used in many parts of Asia, especially Java in Indonesia, a region that, over the years, has excelled in batik production using wax as the principal resist substance. The sub-technique of resist dyeing using wax is known as batik. While this is used as a craft technique worldwide, it is in Java, an island in

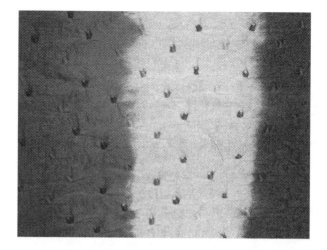

FIGURE 8.48 Detail of tie-and-dye cotton, with knots remaining, mid-twentieth century, Pakistan. Held at University of Leeds.

FIGURE 8.49 Late-twentieth century, West African resist cotton. Held at University of Leeds.

the Indonesian archipelago, where batik reached its highest level of aesthetic excellence. Traditionally, the batiks produced in the sultanates (or palaces, located in cities such as Surakarta and Yogyakarta) were associated closely with Javanese royalty, and differed in terms of colours, motifs and regular patterns used in comparison with batiks produced elsewhere in Java. Further details were provided by Steinmann (1947) and Larsen (1976).

A further possibility was to apply a dye, mordant, discharge or resist paste by simply painting by hand on the cloth's surface. In the modern art-college context, this application may be by brush but, traditionally (in India), a kalam was used; as noted in section 8.2, this is a rudimentary drawing implement used to apply a dye, mordant, resist or discharge paste. In the Indonesian context, an implement known as a canting

FIGURE 8.50 Late-twentieth century, West Africa resist cotton. Held at University of Leeds.

FIGURE 8.51 Late-twentieth century, West Africa resist cotton. Held at University of Leeds.

FIGURE 8.52 Late-twentieth century, West Africa resist cotton. Held at University of Leeds.

FIGURE 8.53 Late-twentieth century, West Africa resist cotton. Held at University of Leeds.

(sometimes spelled tjanting), consisting of a small, tea-pot-like vessel of thin copper, with one or more capillary spouts (through which molten wax flows), attached to a handle generally made from bamboo, was commonly used in the production of wax-resist cloths known as batik. The derivation of the word 'batik' is probably from the Javanese 'ambitik' meaning 'to mark with tiny dots' (Steinmann, 1947). On solidification of the wax, the cloth was immersed in a dye bath, and the dye took to those areas not covered with wax. Wax resists have also been produced using metal blocks (known as cap) to apply the wax (a common practice in Indonesia). In the production of monochromatic batiks, the resist is applied only once, so involves only

FIGURE 8.54   Javanese canting, late-twentieth century, Java. Held at University of Leeds.

FIGURE 8.55   Late-twentieth century, Javanese batik. Held at University of Leeds.

one immersion in the dye bath. Subsequently to drying, resist materials are removed by boiling or scraping. Although the finest Javanese batiks (known as tulis) were produced entirely using the canting and allowed the craftsperson to create any compositional type, regular patterns were produced frequently. A review of Indonesian batik design was provided by Hann (1992).

Traditionally, rice-paste resists were typical of Japan and South China and were used often in association with stencils (known in Japan as katagami) and spatulas. Illustrations from a substantial collection of katagami held at the University of Leeds (UK) were provided by Hann and Moxon (2019: 113–176). Although the bulk of Japanese katagami are free-standing, asymmetrical compositions, a small proportion show regular repetition, and these can be described as regular patterns.

Traditionally, various related techniques were used in West Africa to produce a category of textiles known collectively as adire. These included adire oniko (designs made principally through tying), adire alabare (where designs were produced principally through stitching) and adire elecko (where cassava paste was used as a resist and was applied by hand, possibly using a chicken feather as a brush, or applied through the use of a flat stencil, traditionally made from the sheets of tin lining found in tea chests) (Picton and Mack, 1989: 161–162). The importation of enormous

FIGURE 8.56 Late-twentieth century, Javanese batik. Held at University of Leeds.

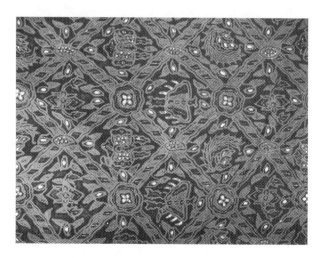

FIGURE 8.57 Late-twentieth century, Javanese batik. Held at University of Leeds.

FIGURE 8.58   Late-twentieth century, Javanese batik. Held at University of Leeds.

FIGURE 8.59   Late-twentieth century, Javanese batik. Held at University of Leeds.

quantities of shirting fabric to West Africa in the early-twentieth century stimulated the development of various textile crafts, particularly this form of resist-dyed cloth. Saheed's publication is a good source for interesting further details (Saheed, 2013). It is worth remarking at this stage that, probably more than any other textile-producing culture historically or in relatively modern times, the Japanese continued to produce astoundingly designed textiles invariably mixing media in ways unfamiliar to non-Japanese cultures. Often resist forms would be combined with forms of printing or weaving. There was also the tendency to mix resist-dyed forms with types of block printing in ways unfamiliar to non-Japanese textile experts who would often puzzle

FIGURE 8.60 Late-twentieth century, Javanese batik. Held at University of Leeds.

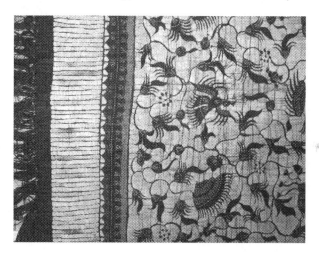

FIGURE 8.61 Late-twentieth century, Indramayu (Indonesia) batik. Held at University of Leeds.

at how a textile was produced, expecting that it could result only from tried and tested combinations known outside Japan.

8.5 IKAT TECHNIQUES

A further resist-dyeing method, known as ikat involves using dye-resistant string (traditionally banana leaf, but in the twenty-first century more likely to be polypropylene) to bind bunches of warp, weft or both series of yarns (to produce warp-, weft- and double-ikat cloths, respectively, each explained further below). The word 'ikat' has origins in the Indonesian (or Malay) word 'mengikat' meaning 'to tie'

FIGURE 8.62   Katagami (mulberry-paper stencil), early-twentieth century, Japan. Held at University of Leeds.

FIGURE 8.63   Japanese tie-and-dye together with block printing. Held at University of Leeds.

(Kramer and Koen, 1987: 265). When immersed in a dye bath, the dye will take only to those areas not covered with the dye-resistant material. Further enhancement through additional dye applications is possible if, after drying, the resist-protected areas are rearranged and dyed again in another dye bath. In warp ikat, only parts of the warp yarns are shielded from the dye. In weft ikat, parts of the weft yarns are shielded from the dye. In double ikat, after shielding, both sets of yarns are dyed and overlap in the final design, so both sets of yarns may together form a motif (Weiner, 1992). A further type of double ikat should be mentioned. This is compound ikat, where both sets of yarns are dyed in parts, but each set of yarns produces an independent motif. A typical visual characteristic of all ikat cloths is a

FIGURE 8.64   South-east Asian cotton ikat. Held at University of Leeds.

FIGURE 8.65   Indonesian warp ikat, with warp yarns resist dyed prior to weaving, late-twentieth century, Indonesia. Held at University of Leeds.

feather-like design effect caused by the dye bleeding beneath the resisting material as well as by the slight movement caused during the weaving process. Sometimes, the clarity of an ikat can be enhanced by ensuring that the embellished yarns are dominant in the final weave; this is especially true with warp and weft ikats (Larsen, 1976: 29). Thus, in a plain-woven warp ikat the warp yarns should be allowed to dominate by being more densely crammed or heavier than the weft yarns, and with a weft-ikat cloth the weft yarns should be given dominance. With double ikat a balanced weave, with a similar density of warp and weft yarns, is most suited. Warming and Gaworski (1981), Gittinger (1985) and Hitchcock (1991) provided substantial reviews and discussion of ikat types associated with Indonesia. Prior to this, Larsen (1976) provided a substantial review of ikats and other forms of resist-dyed cloth. Reviews, focused on Indonesian textile manufacture and covering various forms of resist dyeing, including ikat types, were provided by Hann and Thomson (1993) and Hann (ed.) (2008). Unlike the other resist-dyeing methods, the design aspects of the process are completed prior to weaving. Ikat was common in many parts of Asia (including Cambodia, Myanmar and the Philippines), parts of Africa, Central America and Japan. In Indonesia, the technique was used across numerous islands, including Kalimantan, Sulawesi and Sumatra, with East Sumba producing the most common type of warp ikat (Forshee, 2001). Traditionally, ikats were common in parts of pre-Columbian Central and South America, including Argentina, Bolivia, Ecuador, Guatemala, Mexico, Brazil, Venezuela, Chile and Columbia, and were produced also in Iran as well as parts of Central Asia. Central-Asian warp ikats occasionally depicted repeated arrow-type motifs. These forms were probably produced by arranging the dye-resistant ties in a chequerboard format and, after dyeing and drying, manipulating groups of yarns (into arrow-shaped motifs) prior to weaving.

A cloth known as geringsing, unique to Indonesia, is a double ikat produced only in the Balinese village of Tenganan Pegeringsingan. Typical colours in geringsing cloths are primary red, red/brown, cream, and blue/black. Traditionally, similar techniques have been associated with various producers in India (particularly Gujarat, where the cloth is known as patola) and Japan. An important variety of warp ikat is worth mentioning. This is mashru, a cloth with silk warp and cotton weft, produced occasionally in Gujarat. Importantly, a satin (or warp-faced) weave is used, allowing a dominance of silk on the face of the cloth and cotton on the reverse. Karolia outlined stages of production (2019: 235–244).

8.6 SUMMARY

Hand-carved wooden blocks are clearly an ancient method of applying dyes, mordants, resists or discharge pastes to cloth. These and various related methods were explained briefly in this chapter. Three case studies of traditional block-printed products were presented: ajraks, kalamkari and dabu cloths. The term 'resist dyeing' has been applied to a wide variety of techniques used to embellish textiles by selectively dyeing specific areas on a cloth's surface. Tying, knotting or folding, and the use of stencils or shields, as well as stitching were identified as means of protecting cloths from the penetration of dyestuffs. The application of resist materials such as wax or paste was explained also. Three basic resist-dyeing categories (tie-and-dye, batik and ikat) were identified and product and process variations associated with printing using hand-carved wooden blocks were explained briefly.

REFERENCES

Anon (1967), 'Japanese Resist-dyeing Techniques'. *CIBA Review*, 1967(4):2029.

Barnes, R. (1992), 'Indian Resist-dyed Textiles: The Newberry Collection'. *The Ashmolean*, 22.

Barnes, R. (1997), *Indian Block-Printed Textiles in Egypt. The Newberry Collection in the Ashmolean Museum*, Oxford: Oxford University Press.

Bilgrami, N. (1990), *Sindh Jo Ajrak*, Karachi: Department of Culture and Tourism, Government of Sindh.

Bühler, A. (1951), 'Indigo Dyeing among Primitive Races'. *CIBA Review*, 85:3088–3091.

Bühler, A. (1954), 'Plangi – Tie and Dye Work'. *CIBA Review*, 104:3726–3752.

Burke, K. S., and D. Whitcomb (2004), 'Quṣeir al-Qadīm in the Thirteenth Century: A Community and Its Textiles'. *Ars Orientalis*, 34:82–97.

Else, J. (1988), 'A Composite of Indian Textiles: Tradition and Technology'. *Ars Textrina*, 10:71–84.

Forshee, J. (2001). *Between the Folds: Stories of Cloth, Lives, and Travels from Sumba*, Honolulu: University of Hawai'i Press.

Gittinger, M. (1985). *Splendid Symbols. Textiles and Tradition in Indonesia*, Singapore: Oxford University Press.

Guy, J. (1998). *Woven Cargoes. Indian Textiles in the East*, London: Thames and Hudson.

Haller, R. (1951), 'Indigo'. *CIBA Review*, 85:3072–3087.

Hann, M. A. (1992), 'Unity in Diversity: The Batiks of Java'. *Ars Textrina*, 18:157–170.

Hann, M. A. (2005), 'Patterns of Culture – Techniques of Decoration and Colouration', a monograph in the *Ars Textrina* series, no. 35, Leeds: University of Leeds.

Hann, M. A. (ed.) (2008), 'Patterns of Culture. The Textiles of Bali and Nusa Tenggara', monograph with text mainly by H. Coleman, no. 38 in the Ars *Textrina* series, Leeds: University of Leeds International Textiles Archive in association with the Indonesian Embassy, London.

Hann, M. A., and I. S. Moxon (2019), *Patterns: Design and Composition*, New York: Routledge.

Hann, M. A., and G. M. Thomson (1993). *Unity in Diversity. The Textiles of Indonesia*, Leeds: the University Gallery.

Hitchcock, M. (1991). *Indonesian Textiles*, London: The British Museum Press.

Karolia, A. (2019), *Traditional Indian Handcrafted Textiles*, vol. 1, New Delhi: Niyogi Books.

Kramer, A. L. N., and W. Koen (1987), *Tuttle's Concise Indonesian Dictionary*, Rutland, Vermont, USA: Charles Tuttle Company.

Larsen, J. L. (1976). *The Dyers Art. Ikat, Batik and Plangi*, New York: Van Nostrand Reinhold.

Madan, G. (1990). *India through the Ages*, New Delhi: Publication Division, Ministry of Information and Broadcasting, Government of India.

Murphy, V., and R. Crill (1991), *Tie-dyed Textiles of India: Tradition and Trade*, London: Victoria and Albert Museum.

Picton, J., and J. Mack (1989), *African Textiles*, London: British Museum.

Saheed, Z. S. (2013), 'Adire Textile: A Cultural Heritage and Entrepreneurial Craft in Egbaland, Nigeria', *International Journal of Small Business and Entrepreneurial Research*, 1(1):11–18.

Schaefer, G. (1939), 'The Earliest Specimens of Cloth Printing', *CIBA Review*, 26:914–920.

Southan, M. (2008), *Shibori*, Tunbridge Wells, Kent, UK: Search Press.

Sreenivasam, E. (1989), *The Textiles of India: A Living History*, Ames, Iowa, USA: Octagon Centre for the Arts.

Steinmann, A. (1947). 'Batiks'. *CIBA Review*, 58, July:2090–2123.

Vetterli, W. A. (1951), 'The History of Indigo', *CIBA Review*, 85:3066–3071.

Vogelsang-Eastwood, G. M. (1990), *Resist Dyed Textiles from Quseir al-Qadim*, Egypt, Paris: AEDTA.

Warming, W., and M. Gaworski (1981). *The World of Indonesian Textiles*, London: Serindia Publications.

Weiner, Y. (1992), 'The Indian Origin of Ikat', *Ars Textrina*, 17:57–85.

Wells, K. (1997), *Fabric Dyeing and Printing*, London: Conran Octopus.

Westfall, C. D., and D. Desai (1987), 'Bandhani (Tie Dye)'. *Ars Textrina*, 8:19–28.

9 Finishing

9.1 INTRODUCTION

The aim of textile finishing is to enhance the performance and/or aesthetic characteristics of textile products and their overall perceived quality. Finishes may be durable or non-durable. Durable finishes (such as crease resistance) are not affected by laundering, whereas loss of stiffening, for example, after washing is a case of a non-durable finish that can be renewed through washing, starching and ironing. The concern here is mainly with durable finishes, particularly those of a type which are applied at an industrial rather than at a domestic level.

In general, the term 'textile finishing' is applied to the range of chemical and physical operations, which constitute the final stage of processing prior to cutting the cloth for use. Finishing is carried out on a cloth after it has been produced (through weaving, knitting or some nonwoven method), cleaned, scoured, bleached, dyed and/or printed. The term 'finishing' can also be applied to yarns, such as those intended for sewing or embroidery.

The purpose of this chapter is to identify important stages of processing after fabrication and to explain briefly how these further processes can improve the properties of the textiles in question. Some finishes are specific to fibre type and have evolved with the industry associated with that fibre; these are regarded largely as traditional in nature. A few of the more important of these are reviewed in this chapter. Other processes are more broadly applicable and are invariably used across fibre types. These comprise various mechanical or chemical processes. So, the purpose of this chapter is to identify and explain a range of pre-finishing processes, various finishing processes specific to a few popular fibre types, and to recognise and describe briefly a selection of mechanical and chemical finishing techniques with broad applicability. In this chapter, finishing is regarded as the processes just prior to delivery and cutting for garment or other end use, or just prior to packaging and labelling for the final-stage consumer.

Factors determining the type of finish to be applied to a cloth will include the fibrous composition of the cloth; the anticipated degree of success of the applied finish; the anticipated durability of the finish; the cost of the finish and its application; availability of applicable machinery; environmental considerations, including the release of volatile compounds; and possible unavoidable chemical interactions with other finishes. Other considerations may include costs associated with acquiring the finish chemicals before application, as well as the costs of storage and application. Further concerns may include the degree of stability while in storage; stability of the finish on the cloth during laundering; possible adaptability of the finish to other fibre types; facility for even and consistent application; and biodegradability and environmental friendliness. A good review of these and related requirements was made by Shindler and Hauser (2004). Substantial explanation and discussion, covering numerous aspects of textile finishing, were given by Heywood (2003). Probably the most comprehensive and accessible review of developments in textile finishing across the board, up to the middle of the second decade of the twenty-first century, was provided by Roy Choudhury (2017).

At this stage it is worth underlining the importance of recent developments in finishing. Plasma treatment, where fibrous surfaces are changed, is an important dramatic innovation in textile finishing. Substantial cost savings seem possible, with less water and power consumption involved. Shahidi, Ghoranneviss and Moazzenchi (2014) commented on the environmental credentials of associated innovations: 'Plasma technology, when used effectively, can offer such "greener" possibilities as it is a dry process, is energy efficient, needs a minimum amount of chemicals and there is no down-stream pollution'. Therefore, at the time of writing it seems that plasma finishing offered much potential, with substantial cost saving twinned with eco-friendliness, all the result of relevant research initiated in the late-twentieth century. Often, in the recent past, certain finishes have not been developed and commercialised due to perceived future environmental concerns. Plasma finishing does not appear to be in this category, and at the time of writing there is much ongoing research. Good reviews of various applications were provided by Shishoo (2007) and Morent, De Geyter, Verschuren, De Clerck, Kiekens and Leys (2008).

In the early-twenty-first century, a further development which offered the chance to adjust the properties of textiles in a major way was nanotechnology, where, for example, the addition of tiny (nano-sized) particles to cloth could enhance properties without significant increases in weight, thickness or stiffness. An interesting review article of the possible uses of nanotechnology in a textile context was provided by Kaounides, Yu and Harper (2007).

9.2 PRE-FINISHING PROCESSES

All cloths should be free from impurities prior to finishing, and various pre-finishing processes ensure that this is the case. Textile cloths fresh from the mill are known as grey goods and, before they are ready for application and use, various further stages of processing are required. Singeing (best considered as a pre-finishing treatment) is the process of burning protruding fibre ends on a yarn or cloth to produce a smooth surface, prior to finishing. Starch-type products, known as sizes, used to strengthen warp yarns prior to cloth construction, are washed out from the cloth using a de-sizing agent in some cases and, where the size is water soluble, using hot water. Bleaching assists in the removal of unwanted colouring matter which may be present in the cloth. Scouring is a rigorous cleaning process which aims to remove from the cloth all oils, stains and impurities acquired in processes prior to finishing.

Certain specialised pre-finishing processes are also of relevance to silk and its processing. Foremost among these are degumming and weighting. Degumming is aimed at simply removing the gum surrounding the continuous-filament fibres which make up the cocoon. This is easily achieved by extracting the fibre as the cocoons lie in warm water, though it should be noted that with the various varieties of wild silk it is difficult to remove gum entirely. Occasionally, cultivated-silk yarns are processed without degumming. An example is chiffon (mentioned in section 4.4) which, traditionally, was woven in gummed state with gum removed after weaving. Weighting of silk is aimed at increasing the weight by adding metallic salts at an early extraction stage; however, this process weakens and stiffens the fibre.

9.3   FIBRE-SPECIFIC PROCESSES

Various finishing procedures, specific to each traditional fibre industry, developed over the centuries, focused mainly on improving the physical characteristics of the cloth through the application of the knowledge held at a given time. Often specific techniques and associated machinery would evolve with particular fibres in mind. A few important processes associated with three well-known natural-fibre types (cotton, flax and wool) are outlined in this section.

An important process associated with cotton is mercerisation. Commonly, this process, applied at fibre, yarn or cloth stages, involves soaking the constituent fibres, yarns or cloth in caustic soda. This leads to swelling and change of the twisted-tape shape of cotton fibres (cross-sectionally, under the microscope) to nearly cylindrical rod-like forms, making them smoother, more lustrous, stronger and absorbent, and more ready to take up dyestuffs with brighter and more vibrant colours. Often, cotton sewing thread, which needs to be strong and highly regular, is mercerised.

Flax too has various associated finishing techniques. Probably, a process known as beetling is the most famous of these. Commonly applied to fine linen damasks (and sometimes to cotton equivalents too in order to impart a linen-like quality), the process consists of hammering the cloth across its width (and ultimately down its length) with a series of heavy wooden hammers. This flattens the cloth, improves the lustre and gives a softer handle. Flax, like cotton, can be bleached. In the case of flax, traditionally this was through simply laying the cloth outdoors in grass fields and subjecting it to morning dew and sunlight. With flax, chlorine bleach can be used but care must be exercised, for if the waxes holding fibres together (during growth and subsequently) are damaged then the fibres too will be damaged and weakened (Humphries, 2004: 190).

Unlike cotton and flax, wool is tolerant of acidic conditions but is sensitive to alkalis and chlorine bleach. With wool, scouring at both fibre and cloth stages is necessary. Scouring in the fibre state, prior to spinning, helps to remove soil and vegetable matter as well as suint (perspiration) and wool grease. This is achieved by passing the fibre through a series of bowls, each containing a low concentration of alkali (insufficient concentration to damage the fibre), with a thorough wringing between each bowl. Further residual vegetable matter is removed through carbonisation, a treatment which uses diluted (though heated) sulphuric acid, which chars the vegetable matter and allows it ultimately to be shaken out. Fulling is commonly applied to wool cloth as well. This involves mild surface felting in warm soapy water and enhances the compactness and density of the cloth. Probably, the most important finishing process, unique to wool-type cloths, is mothproofing, where the intention is to make the constituent fibres less digestible to clothes moths and their larvae. Although the principal use was in carpet wool (rather than apparel), a major concern from the late-twentieth century onwards was that all applications to prevent moth attack should be safe to consumers.

9.4 MECHANICAL PROCESSES

Often, finishing treatments are classed as either mechanical or chemical, depending on the method used. Mechanical (or physical) finishing includes various physical finishes, applied using machinery. Although moisture may be used on occasions, largely mechanical finishing is regarded as dry and chemical processing as wet. As observed by Roy Choudhury (2017: 2), often chemical finishing methods are combined with mechanical methods, and whether a finish is classified as mechanical or chemical is based largely on common sense (and answering the question: which makes the most significant contribution?). Roy Choudhury commented: 'the major distinction between the two is what caused the desired … change, the chemical or the machine?' (Roy Choudhury, 2017: 2). With close attention to this distinction, mechanical finishing is introduced below.

Important mechanical finishing processes are aimed mainly at stabilising the cloth; they include applications to improve shrink resistance, to impart permanent pressing, to set the surface shape and to alter the lustre or texture of the cloth. Mechanical finishes are mainly dry processes, involving brushing, shearing, heating, creasing, pressing, embossing, folding or raising. A range of the more common of these is identified and explained briefly below.

Many mechanical processes are based on heat setting, especially where the fibrous content is mainly of thermoplastic fibres. Subjecting such a cloth to heat can help to stabilise or shape it into an intended form, and this may involve permanent pleating, creasing, embossing or puckering. Heat setting can give resistance to wrinkling and crumpling of the cloth. Wires, stiff brushes or emery paper, each in roller form, can be applied to a cloth to roughen or raise the cloth's surface and thus its texture. Shearing using a blade of some kind is aimed at removing surface fibres on a cloth. Much mechanical finishing reduces stresses on the cloth through compressing it, and thus lessens the possibility of further shrinkage during future laundering.

Heat and/or pressure can be applied to cloths in order to improve their performance or appearance. Mechanical finishes include calendering, embossing and napping. Calendering involves passing cloth (full width), between heated pressure rollers, flattening the yarn, making the cloth softer and increasing its lustre. In the process known as frictional calendering a speed differential operates between the two rollers to impart a shiny, glossy, chintz-type surface finish. Embossing employs heat and pressure by passing cloth (typically a nonwoven cloth) between a heated engraved roller and a roller of softer material. Commonly, embossing is on light-weight cloth where a logo of some kind is imparted. Embossing can give a seeming raised-surface effect, caused by passing a cloth under pressure between heated rollers, each with an embossed surface. With napping, traditionally applied to woollen cloth, but more widely applied to other cloth types by the mid-twentieth century, the fibre is raised by passing the cloth between rollers covered by hooked metallic needles. Examples include velvet- and flannelette-type cloths. A similar raised result is what is known as peach-skin finish applied to finely woven microfibre cloths. Brushing can remove loose fibres from a cloth's surface, but the term 'brushed' cloth is used commonly where surface fibres are brushed or raised to give a nap surface with a soft handle. Sanforising, used mainly on cotton cloths, is a controlled mechanical shrinkage, without chemical applications, involving passing the cloth with the addition of steam between a series of belts and cylinders. Flocking (or flock printing, mentioned previously in section 7.4) is a process which involves either applying an adhesive to parts of a cloth's surface and sprinkling short fibres over those parts or allowing the short fibres to adhere to the cloth's surface through imposing an electrostatic charge; the former could be considered as a chemical finishing process and the latter as a mechanical finishing process. Shearing (also referred to as cropping) is a cutting process aimed at removing extra protruding fibres from a cloth's surface. During cloth formation, high levels of stress are applied, and these build up and are held in the cloth. Compacting aims to accommodate these inbuilt stresses by releasing them in a controlled way and ensuring a stable or compact structure for future use. Compacting permits the cutting and application of the cloth to an end use without the possibility of further shrinkage during subsequent laundering. The sanforising process (mentioned above) is an example of a compacting procedure.

An important machine in most industrial-level dyeing and finishing environments is the stenter (or tenter, the name given commonly to a similar machine used in the drying of wool cloth), a machine which holds the cloth in open-width format on a series of pins through each selvedge. The stenter is of value in the fixing of finishing agents, in heat setting, curing and in the application of a form of mechanical finishing known as stenter finishing; further details were given by Roy Choudhury (2017: 16–18).

9.5   CHEMICAL PROCESSES

Most chemical finishes are applied using machinery dedicated in the first instance to textile dyeing. Probably, the most popular means of application is using a padding mangle, where the cloth is immersed in a finishing liquid, squeezed between two rollers, dried and then cured in another machine. The process has been referred to often as pad-dry-cure. A good, brief, review was given by Roy Choudhury (2017: 7–13).

An important challenge in chemical finishing is ensuring that constituent finishes (which need to be combined in one container for reasons of economy) are compatible. Some components may indeed assist each other, but others may conflict (Roy Choudhury, 2017: 6). Compatibility of constituent chemicals to each other as well as to the fibre content is an important consideration. The environmental characteristics and sustainable nature of finishes proved also to be crucial considerations by the early-twenty-first century. Much useful information was presented in Blackburn (ed.) (2009).

Various resin finishes are intended to stabilise the fibres in some way, possibly making them resistant to water, soiling, fire, creases or shrinkage. Water-repellent finishes (different from water proofing) ensure that the cloth can withstand a brief shower. Anti-soiling finishes repel stains or adjust the cloth's surface to aid the removal of impurities during laundering. Fire-retardant finishes can reduce the tendency of a cloth to burn, or to propagate the flame. Some fire retardants may be durable, and others may be non-durable. Problems associated with the application of fire retardants include unpleasant odours, yellowing, colour change, cloth stiffening, possible skin irritation and loss of the cloth's tensile strength.

Various terms are used in the assessment of a cloth's performance, particularly its resistance to water. Three terms appear to dominate: 'water repellence', 'water resistance' and 'water proofing'. For a cloth to be water repellent, it should not be penetrated easily by water and has probably been coated with a barrier to prevent water penetration. Water-repellence applications may involve coating a cloth with a solid barrier such as polyurethane, but this may lead to discomfort in apparel as it prevents the passage of water vapour in the other direction. Water resistance is when a cloth can resist the penetration of water to a certain degree but not totally. Water proofing is when a cloth is totally impervious to the penetration of water. Waterproof cloths are coated with rubber, waxes, oils or plastic resins and these prevent the penetration of water. A waterproof finish makes a cloth resistant to water penetration, often by simply applying one or more coatings of polyurethane or similar additions, invariably aimed at uses such as cyclewear, rainwear or sea wear. With respect to garments, the aim is a breathable cloth which resists water, on the one hand, but ensures the comfort of the wearer, on the other, by allowing the passage of water vapour away from the body within the garment. Water-repellent finishes, although not fully waterproof, can give greater comfort in use, provided the aim of breathability is acknowledged.

Cloths from both natural and manufactured fibres attract dirt, with manufactured fibres probably the worse of the two (as the constituent fibres are probably more hesitant in releasing dirt during laundering). A soil-release finish, for both fibre types, will not prevent dirt from attaching to a cloth but may allow it to detach more easily. Dirt will be entrapped in the spaces between fibres and yarns (so the type of fibre and yarn construction can influence a cloth's tendency to soil). Dirt which is oily in nature is more difficult to remove as it can penetrate the fibres. Where static electricity builds up (probably the case with many manufactured fibres), dust particles are attracted. Soil-release finishes are effective for cloths from both natural and manufactured fibres.

Flame-retardant cloths will resist fire though they are not fully fireproof. A flame-retardant addition will act to suppress, delay or reduce combustion and the propagation of further flames. This is of great importance across a wide range of products in addition to those cloths destined for personal-apparel use. Where cellulosic fibres are the principal fibre constituent, a loss of strength and a harshening of handle may be among the drawbacks of application. A major concern is the tendency of synthetic fibres, such as nylon and polyester (both of relatively low flammability), to cause serious injury by melting and dripping. Flame-retardant additions are, however, possible in these cases.

Anti-microbial finishes can protect users from odour-causing micro-organisms on the cloth and can inhibit bacterial and fungal attack on the cloth. In the early-twenty-first century, silver-based finishes offered much respite but had various associated environmental concerns, so alternative anti-microbial additions received attention; the situation by the end of the second decade of the twenty-first century was summarised by Hebden and Goswami (2018: 365–366). Anti-bacterial (or anti-microbial) finishes give protection against the propagation of odour-causing bacteria (which can lead to staining and other detrimental effects).

Anti-shrink additions may reduce cloth shrinkage. In addition to resin applications, further processes may include steaming and machine shrinking. Probably, the universal textile finish is pre-shrinking, as most cloths will shrink when laundered for the first time. Untreated cotton cloths, for example, tend to crease and shrink and easy-care finishes can keep this tendency to a minimum. Across all cloths, a degree of dimensional stability is desired, and this is often given by a pad-dry-cure addition, where a cross-linking chemical (which helps to hold internal structures in place) together with a catalyst (to encourage cross-linking) and other desirable components are dried into the cloth and then cured. Further finishes e.g. to enhance softening, fire retardation or stain repellence may be added also prior to drying. A good review of finishing possibilities was given by Hebden and Goswami (2018: 366–367).

Pilling is the formation of entangled fibres (often in spherical format) on the surface of a cloth, spoiling the appearance. Fibres with a high tenacity (such as nylon and polyester) are particularly prone to pilling, but with most natural fibres pills simply fall off during wear. Anti-pilling finishes, either through mechanical or through chemical attention, have been made available.

By the late-twentieth century, the potential dangers offered by ultraviolet (UV) radiation were broadly recognised globally and this led to the desire to enhance textile cloths, where it was believed that they might be worn in strong sunlight (including sun hats, T-shirts and swimwear), through the addition of UV protection of some kind. Often UV-absorbent finishes were applied at the same time as dye application. UV or sunlight resistance minimises the degradation effect of UV rays on colour and fibres. Many manufactured fibres build up static charges which can be eliminated by including an antistatic agent in the polymer during manufacture. Antistatic finishes can prevent static build-up on a cloth.

In order to encourage improved handle, and thus enhance the perceived quality of a cloth, softeners of various kinds may be added, though often these are removed through repeated laundering. Traditionally, early forms attempted to emulate natural softeners such as lanolin (associated closely with sheep's wool). The introduction and use of microfibres led to an increased awareness among consumers of cloth handle as an important attribute. A brief, though well-focused, review of possibilities was given by Hebden and Goswami (2018: 367–370).

9.6 SUMMARY

This chapter has been concerned with the final stage of processing of cloths prior to making up into garments or other products. It was recognised that textile finishing prepared cloths further for everyday use. A selection of processes was introduced, including a few carried out at a preliminary stage, a few fibre-specific finishes such as beetling and mercerisation, and various processes applicable across fibre types. All were aimed at ensuring that the finished cloth was fit for its intended purpose. Plasma finishing was recognised as an important future development, due largely to its favourable environmental attributes.

REFERENCES

Blackburn R. S. (ed.) (2009), *Sustainable Textiles*, Cambridge: Woodhead.

Hebden, A. J., and P. Goswami (2018), 'Textile Finishing'. in T. Cassidy and P. Goswami, *Textile and Clothing Design Technology*, London and New York: CRC Press (Taylor and Francis group), pp. 357–373.

Heywood, D. (ed.) (2003), *Textile Finishing*, Bradford: Society of Dyers and Colourists.

Humphries, M. (2004), *Fabric Reference*, Upper Saddle River, NJ: Pearson.

Kaounides, L., H. Yu, and T. Harper (2007), 'Nanotechnology Innovation and Application in Textiles Industry: Current Markets and Future Trends'. *Materials Technology*, 22 (4):209–237.

Morent, R., N. De Geyter, J. Verschuren, K. De Clerck, P. Kiekrns and C. Leys (2008), 'Non-thermal Plasma Treatment of Textiles'. *Surface Coating Technology*, 202:3427–3449.

Roy Choudhury, A. K. (2017), *Principles of Textile Finishing*, Cambridge, UK: Woodhead.

Shahidi, S., M. Ghoranneviss and B. Moazzenchi (2014), 'New Advances in Plasma Technology for Textile'. *Journal of Fusion Energy*, 33:97–102.

Shindler, W. D., and P. J. Hauser (2004), *Chemical Finishing of Textiles*, Cambridge, UK: Woodhead.

Shishoo, R. (ed.) (2007), *Plasma Technologies for Textiles*, Cambridge, UK: Woodhead.

10 Motifs and patterns

10.1 INTRODUCTION

When considering textile design worldwide, the important influence of trade is immediately apparent. For example, in the twentieth century, East African kanga cloths showed influences from India, China and Europe. A design type which became known widely as the Paisley design, and had close associations with shawl-making centres in Britain and France in the late-eighteenth and early-nineteenth centuries, had origins in the Indian province of Kashmir and, before this, was probably derived from Persian floral designs. Early initiation of trade between Europe and China was famously along the silk route, involving numerous cultures and nations along the way, and many of the motifs and patterns on textiles and other embellished surfaces used throughout Europe and Asia for much of the second millennium CE were deeply influenced by examples from this source.

Often the nature of regular-pattern design was determined by technological change and diffusion as well. Initially in Europe, basic handloom weaving and early forms of printing were displaced by various technological innovations during the Industrial Revolution. A process of technological change, innovation, adoption and adaptation was followed by a worldwide focus on increasing mechanisation and lowering labour costs. In parallel, during the late-nineteenth and early-twentieth centuries, there was a rise in a category of creative individuals, known as designers, who were capable of working within the constraints of the newer processing technology. In the main, these constraints were driven by economic factors such as restrictions associated with ensuring that motifs were rendered with the minimum complexity and in a limited number of colours. This colour constraint was often a major consideration in the pre-digital printing era, as it was likely that a separate block, plate, roller or screen would be required for each colour and this need often incurred much expense.

The engagement of designers and the associated high labour- and skill-intensive approach to regular-pattern design was, by the early-twenty-first century, supplemented, though not replaced, by various computer-aided design techniques. Importantly, in printed textile design where digital forms of printing were in use, the restrictions on permitted numbers of colours and the strict repetition of repeating units were loosened. However, the essential early considerations relating to anticipated end use, design theme, selection of raw materials, texture, line and form were retained and were still (by the early-twenty-first century) crucial considerations at the preliminary stages of the design process. It is worth mentioning that, although the restrictions on the number of colours permitted were removed in contexts where digital printing was adopted, fashionable colour palettes (of up to around ten colours) were still the market preference. While the role of the textile designer had changed to a small degree with the necessity to keep abreast of digital changes, there remained the need for creative individuals to ascertain what the market required and how best to meet that requirement with the facilities and resources at hand.

The objectives of this chapter are thus to: identify important motifs, patterns and thematic types; explain the use of grids in design development; explain that the anticipated means of manufacture and end use are both of importance throughout the design process; underline the fact that different means of production place different restrictions on the nature of final designs; and present an outline of the nature of design traditionally associated with different geographical regions.

10.2   MOTIFS, PATTERNS AND THEMATIC TYPES

Textile designers must develop an awareness of motifs and patterns. Motifs may be of various historical or cultural origins, or themes or scales. It is important that designers acknowledge the means of manufacture while keeping their minds firmly focused on the anticipated end use. Motif types can be drawn from nature or the urban environment, and may be of floral or plant origins, of animal or landscape origins, architectural, mechanical, electronic objects or products, as well as geometric figures, often including straight lines and sharp angles, often representing two dimensions but occasionally three dimensions, as well as spots, stripes and checks, sometimes on their own and sometimes in combination. When used in the textile context, the term 'patterns' is generally used to refer to those designs that show regular repetition of a unit of some kind (frequently referred to as a motif). A comprehensive treatise was provided previously by Hann (2019).

Often, historical or antique designs serve as a starting point for the development of design themes and are re-interpreted by experienced designers, whose intentions may be towards future fashion or interior textiles, wallpapers, wrapping papers or some other design area. Design themes may spring from what appears to be infinite possibilities relating to human cultural endeavour, life or the environment. They may refer to different time periods, art movements or techniques (e.g. Victorian or Georgian, Impressionism or Art Deco, ikat or block-printed cottons, respectively). Often themes such as these gain the attention of publishers, museums or galleries, and invariably illustrated publications or catalogues result. Also, designers will often be aware of changing interests and will visit galleries, museums and archives. Design themes come and go, and like much else in early-twenty-first-century life are subject to changing fashions and the time span of their dominance is unknown in advance. Briggs-Goode and Townsend (2011) provided a comprehensive account of the fundamentals of composition and repetition, with a focus on aspects of printed-textile design.

10.3   GRIDS

The bulk of textiles produced by industrial means show regular repetition, and in order to ensure that repeats are created to the correct size and orientation and at the correct distance apart, often a regular-square grid structure is used, and copies of the repeated component placed within this structure. Hann (2015) focused attention on how regularly repeating patterns were often built on underlying grid structures, though Hann and Moxon (2019) and Hann (2019) explained further possibilities, and in particular how an underlying grid could be manipulated as a compositional structure and how constituent cells within grids could be stretched in one or more directions.

10.4   MEANS OF PRODUCTION

As the twentieth century progressed, regenerated cellulosic and other manufactured fibres of various kinds gained popularity alongside numerous technological innovations focused largely on novel, cheaper and faster means of yarn and cloth processing. By the end of the twentieth century, textile and clothing manufacture as industrial activities had migrated largely from European countries to China and elsewhere in Asia. A decade-by-decade breakdown of twentieth-century influences, changes and developments associated with textile and clothing manufacture was given by Udale (2014: 14–21). Probably the most far-reaching technological influence had been the widespread acceptance of computerisation and digital control of manufacture. An increased awareness of the environment and of the contribution made by textile and clothing manufacture to its detriment was a key development worldwide in the early-twenty-first century also. Recycling and upcycling of used products were key developments, and consumers became increasingly aware of the lack of durability and lack of sustainability associated with textile and clothing manufacture.

It is assumed often by observers with little textile knowledge that all designs can be created by any means, using, for example, printing, knitting or weaving. This is clearly not the case and, when developing a suitable design, designers must take account of the manufacturing technique available, though it should be recognised that digital printing has revolutionised textile printing, with restrictions on number of colours and area of repeating unit largely being problems from the pre-digital age.

Worbin (2010) focused on the ability of digital technology to be integrated into the substance of textiles and to extend their function and commented that:

> Textiles are traditionally designed and produced to keep a given, static expression during their life cycle; a striped pattern is supposed to keep its stripes. In the same way textile designers are trained to design for static expressions, where patterns and decorations are meant to last in a specific manner. However, things are changing. The textile designer now deals also with a new raw material: a dynamic textile, ready to be further designed, developed and/or programmed depending on functional context.
>
> **(Worbin, 2010, thesis abstract, page not numbered)**

A novel view was taken also by Murray and Winteringham (2015), in their magnificently illustrated and refreshing treatise *Patternity*, where they argued that patterns (interpreted in a wide sense to include a vast range of visual statements) could offer new ways of seeing and understanding the world around us.

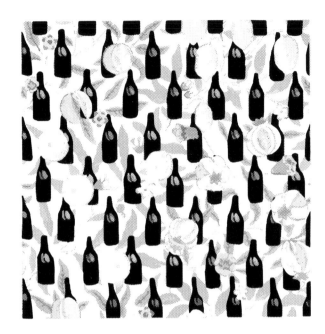

FIGURE 10.1A–H   Rich sources of design ideas are offered to designers on consideration of historical pieces held in museums and galleries. The illustrations included here have been drawn (JZ) from historical pieces held at the Victoria and Albert Museum (London). Each is shown in black and white only. Some show obvious repetition and some not. The use of computing technology allows for the easy repetition of component parts, should the designer desire to do so. Designers often collect such images together with associated palettes and later colour these outlines in a manner which they believe will be commercial in the future. It may well be the case that a collection has been sourced from nature, modern architecture and historic textile designs, with some aspects coming from each source.

10.5   WORLDWIDE DISTRIBUTION

Concentrating on the European context, Udale provided a concise, though well-focused, review of historical developments of relevance to changes in textile manufacture (2014: 11–15). The profound influence from other cultural contexts is apparent. Some of the main changes are summarised below.

During the seventeenth century, the French government supported the then-evolving silk industry of Lyons which, after the application of technological innovations in weaving and dyeing, produced higher qualities of silk than were typical of Italian producers who had held commercial sway in the previous century. Seventeenth-century Rococo visual arts influenced embellishments on textiles. Of importance also were influences from cultures to the east of Europe, especially China and Japan, collectively influencing the production of items lumped under the general title of Chinoiserie. From sources such as these, unfamiliar colour combinations, extracted particularly from kimonos imported from Japan, were dominant.

FIGURE 10.1B Continued.

In the early 1700s, exotic plant and floral silk designs showed a strong influence from Asia and inspired producers in Europe. Known as bizarre silks, they laid the basis for the widespread acceptance of lace- and floral-type motifs by the third decade of the century. A further major influence on manufacture in Europe were the hand-printed cottons from India known commonly as chintz. These provided relatively cheap, bright and colourfast cottons to the extent that both French and British manufacturers came under extreme competition from these superior products and this, in turn, led to the introduction in each case of restrictive legislation, banning the import and use of these Indian textiles. Relevant legislation was repealed in the latter half of the eighteenth century, and this repeal seemed to have a positive influence on industrial development in Europe, especially France (Udale, 2014: 12). By the late 1700s, shawls of cashmere wool were imported into Europe from north-west India and this helped to stimulate European technological innovations further. Some manufacturers, particularly in France and Britain, produced shawls in imitation of those imported from India; by the late 1700s probably the most noted centre was Paisley in Scotland, which gave its name to the characteristic tear-shaped motif.

FIGURE 10.1C Continued.

By the early 1800s, small-scale floral designs were popular throughout much of Europe. Key technological innovations, such as the Jacquard loom and various mechanised forms of printing, were widely accepted in both France and Britain. In Britain, the perceived deterioration in good design stimulated developments associated with the Arts and Crafts movement, led by individuals such as William Morris. Towards the end of the nineteenth century a strong Japanese influence continued to be apparent, readily seen in Britain through the work of pattern designers such as Lewis Foreman Day. A notable publication, which continued to influence pattern designers in subsequent years, was Owen Jones's *The Grammar of Ornament*, a treatise first published in the mid-nineteenth century and with numerous further editions (Jones, 1856). In the early-twentieth century there was an increased interest in ancient Egypt and this coincided with the developing Art Deco Movement.

It was certainly the case that design types were associated with certain geographical regions and historical periods. Different parts of the world have specialised in

FIGURE 10.1D Continued.

certain types of raw material, processes, products, motifs and patterns depending on numerous issues, including the environment, trade, skills and availability of raw materials. Symbolism is often a feature, with motifs or patterns interpreted differently across cultures and sometimes even within a culture. Often a particular raw material or means of production is connected closely with a particular historical or cultural context. Ancient China is associated closely with silk processing, complex woven silks produced on draw looms and the production of particularly fine embroidery. Indonesia (from the middle of the second millennium CE onwards) is perceived as a place for batik and ikat manufacture. Central Asia (historically and in relatively modern times) is regarded as a location for wool-felt manufacture as well as large-scale warp ikats in silk with a dominance of arrow-type designs. Traditionally, Turkey and Iran have been connected to hand-tufted carpet manufacture, and ancient Egypt with flax processing and linen manufacture. West Africa, especially Nigeria, is regarded as a location for indigo-coloured, resist-dyed cloths. South Asia, particularly India, is associated with hand-carved wood-block printing and resist-dyed cotton cloths. There are numerous further examples. By the beginning of the third decade of the twenty-first century, visual records of a historical or geographical nature remained of importance to many designers, possibly simply as a source of inspiration.

FIGURE 10.1E Continued.

FIGURE 10.1F Continued.

Often knowledge and popularity spring from research and discovery. Knowledge of production and trade in Indian textiles developed substantially in the wake of research and exhibitions, particularly in locations such as the Victoria and Albert Museum. From the viewpoint of designers keen to develop their knowledge of textile design in different cultures, and thus inspire their own original design work,

FIGURE 10.1G Continued.

locations such as the Victoria and Albert Museum and museums worldwide have been particularly rich sources, with numerous publications being produced over the years. Some of the relevant literature associated with the Victoria and Albert Museum and elsewhere is identified below.

Wilhide (ed.) (2018) explored repetition in regular patterns from a historical perspective and made connections between this and observations in the natural world (including the petals of flowers, spirals in snail shells and growth in stems and vines). Jackson (2018) presented a readily understandable treatise, avoiding advanced mathematical terminology, on how motifs are brought together in the creation of patterns. Adopting a symmetry focus, Jackson (2018) explained and illustrated how all regular repeating patterns are based on the four symmetry operations of translation, reflection, rotation and glide reflection (see also Hann, 2012: 72–105).

There appears to be a tendency (principally from individuals with no real design training, background or appreciation) to believe that skill, knowledge and experience are not required in the development of original designs. Rather there appears to be the belief among some that it is simply a matter of taking historical designs (no

FIGURE 10.1H   Continued.

matter what the intended end use) and reproducing these on fresh cloth in the same colours and to the exact scale as previously. The adaptation of traditional motifs and patterns is fine, but such adaptation must focus on the precise anticipated end use, the intended colour palette and anticipated market. It is often best to integrate past concepts with new visual achievements (possibly obtained through further observation and drawing). So, all in all, it is wrong to assume that the design journey is akin to a process of painting by numbers.

10.6   SUMMARY

Trade in textiles historically was an important influence on the nature of motifs and patterns used on textile cloths. Numerous motifs, sourced from the natural and urban environment, as well as various abstract signs or symbols, have been used to embellish textiles both before and after the Industrial Revolution. Some cloth types, especially stripes and checks, may well have been inspired by the associated technology used, where stripes may have been suggested by simply seeing alternative coloured yarns placed side by side in a warp- or weft-ways direction, with more complicated forms as well as checked-type embellishments being an obvious further progression.

Grids formed from square unit cells are an obvious means of measuring and estimating motif size and its repetition. From the viewpoint of design, probably the most far-reaching innovation since the Industrial Revolution was the introduction of computer software which allowed textile designers greater creative freedom than ever before. Digital printing did not necessitate the use of only a small number of colours within a tightly specified repeat area. Rather, designers could use as many colours as they wished and could choose whatever depth of repeat they wished.

REFERENCES

Briggs-Goode, A., and K. Townsend (eds.) (2011), *Textile Design. Principles, Advances and Applications*, Cambridge, UK: Woodhead.

Hann, M. A. (2012), *Structure and Form in Design*, London: Berg.

Hann, M. A. (2015), *Stripes, Grids and Checks*, London: Bloomsbury.

Hann, M. A. (2019), *The Grammar of Pattern*, London: CRC Press (Taylor and Francis group).

Hann, M. A., and I. S. Moxon (2019), *Patterns: Design and Composition*, New York: Routledge.

Jackson, P. (2018), *How to Make Repeat Patterns*, London: Laurence King Publishing.

Jones, O. (1856), *The Grammar of Ornament*, London: Day and Son.

Murray, A., and G. Winteringham (2015), *Patternity. A New Way of Seeing*, London: Conran Octopus.

Udale, J. (2014). *Textiles and Fashion*, second edition, London and New York: Bloomsbury.

Wilhide, E. (ed.) (2018), *Pattern Design*, London: Thames and Hudson.

Worbin, L. (2010), 'Designing Dynamic Textile Patterns', unpublished doctoral dissertation, Chalmers University of Technology, Gothenburg (Sweden).

11 Design development and presentation

11.1 INTRODUCTION

The intention of this chapter is to stimulate an awareness that design is a process, requiring much preparation and focus. Although computing technology helps to lessen the tedium of design (such as, for example, ensuring exact repetition of repeating units, if that is the intention of the designer), there is still a necessity for trainee designers to develop their aesthetic awareness through observation, representation, visual research and development. The emphasis throughout much of the present book is on products, machinery and equipment used in the manufacture of textiles, but the focus in this chapter is quite different. Here the intention is to identify the stages in the preparation of designs for consideration prior to possible manufacture.

11.2 MARKET RESEARCH

The term 'textile design' refers to the process of creation aimed at a specific end use, employing knitting, weaving, printing, or other means or combinations. Where a textile collection may be intended for a specified consumer or group, the design process may begin with a consideration of market aspects of the collection as well as discussion of the precise needs of consumers and an identification of the desirable attributes of the final collection. Ideas at an early stage may involve information relating to anticipated trends, raw materials, market or associated branding, if applicable.

Textile products may be aimed at a variety of markets and end uses, including clothing (often with a fashion emphasis), architectural interior uses and a wide range of what are often referred to as technical end uses (a term which seems to cover medical, outdoor architectural, automobile or general transport, marine, aeronautical or landscape contexts). Sometimes the term 'technical textiles' is used where fibres have helped to impart various desirable characteristics when used in association with non-textile raw materials such as plastics and concrete. The inclusion of fibres in various composite forms was a particularly important development in the automobile industry, for example. However, by the early-twenty-first century it appeared that a substantial proportion of the worldwide market for fibrous articles was in clothing end uses (and this destination has been the principal focus within this present book).

Often the creative stages, from preliminary concepts to the production of the final design collection, are many and, frequently, begin with a process of identifying the relevant market and gathering visual information (an outlook common to enquiry associated with all visual-arts-and-design disciplines). Market research and gathering of ideas are crucial foundations for any worthwhile collection.

Where the design process is focused on the very specific needs of a single consumer, then a one-off customisation can be seen to be of relevance. However, this is rarely the case and normally designers will focus on the needs of a mass market. Technological aspects of manufacture, including the addressing of various sustainability and environmental concerns, were typical considerations during the second decade of the twenty-first century.

Sources of raw materials may need to be known (to determine whether they are ethical in nature). Questions may include: What is the anticipated life cycle of the product, and can the product be recycled? Environmental and ethical approaches taken by some companies may well be announced as part of their marketing strategy.

Fashion forecasting involves identifying styles and colours which will be considered desirable by consumers in the future, providing such information a few years in advance of manufacture to ensure ample production time throughout the supply chain. Frequently, textile designers work closely with design briefs inspired by these forecasts. By the early-twenty-first century, it was well known that colour prediction was closely related to clothing-fashion forecasting, but its value extended to other industries also (often with different colour palettes required) including, for example, automobiles, domestic and office interiors, mobile phones and other consumer products; in these areas colour change had developed into a dominant force governing future market success or failure.

By the early-twenty-first century, fashion forecasting and colour-palette prediction were rarely done in-house but rather were provided by specialist companies focused on key market themes, styles and colour palettes anticipated for use in different product areas up to three years in advance. This time frame of up to three years was the norm and this allowed yarn manufacturers, fabricators and dyers to equip the industry to meet anticipated consumer demand. Colour and style prediction were not just crucial to textile manufacturers aiming at fashion end uses, but to manufacturers across the full range of consumer products from mobile phones to automobiles, packaging and interiors. By the early-twenty-first century, different products continued to have different frequencies for anticipated changes in demand, but in the textile context the Spring/Summer and Autumn/ Winter collections continued to dominate; even within the context of so-called fast fashion as a dominant force, palettes and styles remained aligned closely to those anticipated within the two seasons for a particular year, though minor style additions during the year (in the form of what became known as capsule collections) may have been added.

11.3   SOURCES

To create a successful collection of designs, it is necessary for the designer to refer to numerous sources, including information relating to market trends, especially colour palettes and anticipated textural qualities, as well as manufacturing circumstances and budget constraints. Markets need to be researched, forecasting websites need to be considered, influences on consumer preferences (including considerations such as percentage rates of interest and inflation, and anticipated economic conditions in general) need to be accounted for and a theme board (known also as a story or mood board), giving an indication of sources of visual development, anticipated colour palette, various production-related information and end use, needs to be developed to accompany each design collection. Often, a design collection with accompanying information on the theme board will, first, help to convince the manufacturer of the worth of a collection and, second, may influence decisions to adopt the collection for production. Further relevant remarks relating to theme boards are made in section 11.5. Knowledge of the intended end use is crucial to the designer as this can help to focus design ideas. In all cases, the designer will not just attempt to meet the client's specifications but will try also to ensure that the intended product is in some way superior to what will be available from the client's competitors.

As with other visual-arts-and-design disciplines, the creation of a collection of successful textile designs requires research and development, in parallel with the consideration of a wide range of aesthetic, technological, cost and market concerns. Sources for visual research and development are often referred to as inspiration, and these may vary from consideration of aspects of the natural or rural environment to the urban and built environment; inspiration may be at the micro or macro level, involving, for example, colours, textures, structures, sounds, smells and temperatures. Often, in the formal academic environment inspirational sources are classed as primary or secondary in nature, depending on whether the source has been sketched, painted, photographed or reproduced in some other way by the designer (primary research), or has been created, produced, published, manufactured or designed by others (secondary research). Primary research sources include sketch books, drawings and original photographs taken by the designer. Textile designers may collect objects (such as, for example, pieces of driftwood found on a seashore) which they find aesthetically appealing. Items such as these are of value because they may have a certain colour, weight, shape or texture, and may help to recall a favourable experience at a time associated with their collection. This and associated images are sources unique to the designer and, as such, constitute primary visual research sources. With secondary research sources, often images are sourced from magazines, books, films or physical objects, and are gathered and collected from commercial or other sources. Typically, the designer creates a collage of images from both primary and secondary sources, with adjacent images having a colour, thematic or textural relationship; this collage may then form the basis of the theme board (which may also express various market, cost and production considerations). Collages and theme boards can be created digitally but are often presented to the customer in physical form.

Referring to types of primary source, Steed and Stevenson (2012: 37) observed that these may include 'conversations or interviews with other people. Listening to someone tell a story, for example, may conjure up moods and colours or textures and feelings and … may also include festivals or carnivals … which may evoke moods of humour, exuberance, sobriety or decadence'. A further, revised edition of Steed and Stevenson (2020) is worth seeking out. So, primary sources may not necessarily be purely visual but, rather, may relate to an interaction of some kind between the designer and a physical environment or with another individual or group of individuals, and may include sounds, tastes and temperature, each of which may be reminiscent of a visual quality of some kind. Secondary research, seen initially through the eyes of someone else, may include images or other ideas from periodicals, exhibition catalogues, magazines, websites and books, and museum, gallery and archive visits. Often the outcome may be a blend of primary and secondary sources, with a spark of individuality abstracted from primary sources, yet with an assurance of acceptability by the inclusion of secondary sources.

11.4 OBSERVATION, VISUALISATION AND REPRESENTATION

Observational drawing is a fundamental aspect of design sourcing and development. The drawing techniques employed are aimed at capturing the essence and not necessarily the true or realistic representation associated with proportion and relative scale. Drawing may employ a wide range of implements or materials including pencils, charcoal, crayon or pastel, gouache, tape, ink or brushes of various kinds. The techniques employed may include the use of inks, scratching, bleaching, sponging, stitching, scraping, spraying, or collage on stretched or other sorts of paper. Considerations relating to scale, colour and texture are all of importance. The use of masking fluid, wax, crayon or application of a wash may be of importance also. Using ink or pens on a wet surface will allow the ink to bleed or diffuse and spread. Designers can learn control through experimentation and experience, by changing the surface and using different weights of cartridge or watercolour paper. Gouache and other sorts of watercolours are often used by designers as they are mixed easily, dry quickly and hold the colour well. Sponges may be used to apply such paint. Scratching of the painted surface may be by means of a stick, scalpel, pencil, nail, pin or penknife blade. The use of a sewing machine for outlines or hand sewing to pull the thread to create ruffles in the cloth are further possibilities.

Mixed media use two or more techniques (combined in one composition). This may show a combination of traditional techniques and surfaces used with, for example, collage, relief drawing, paper cutting, folding and tearing. Shredding, bending, twisting, crumpling, etc., may be of importance also. Changes in scale from very tiny to very large can be used to good effect to explore the characteristics of a subject (e.g. small-scale insects in relation to large-scale insects). When research has been completed, images often need to be presented to a client before designs can evolve further and reach completion. An editing process is required, and this is best left to experienced designers.

The preferred working environment for most designers is a studio, a space which facilitates creative experimentation and achievement. Large companies may boast that they use their studio as the creative hub of their activities. Designers work in a variety of ways, adopting different procedures and following personal preferences with respect to colour palettes, themes and media. With the passage of time, all visual artists and designers build up collections of favoured equipment to assist with analysis and development. Some may prefer particular types of pens, pencils, crayons, charcoal sticks, gouache or other paints, cartridge or tissue paper of a given weight, small, medium or large sized, bleached white or toned, squared or divided in some other manner. Therefore, the best advice with respect to equipment is for the designer to select what seems most comfortable as time goes by. In setting up a studio, however, certain essentials seem appropriate; the more important of these are identified below.

Successful collections of textile designs are often based on visual research and development from a selection of inspirational sources. This process of research and development is a process of visualisation in which the designer abstracts or edits desirable aesthetic aspects from the selected sources. To do this, the designer observes sources and represents ideas through drawing or similar means of recording. In fact, observational drawing forms an important skill expected from all visual artists and designers. The intention is to record visually stimulating sources and to use this information as a source for further visual development and, ultimately, the production of a design collection which captures the mood of the original selected sources. Designers need first to have various visual recording equipment at hand; secondly, they need to refer to a selection of visual and/or other sources; thirdly, they need to record visually favourable aspects of these sources and finally incorporate these aspects into a design collection. Thus, in the preparation of a collection, designers must undertake visual research. Often, they must observe and analyse using a range of equipment held within a studio environment. It is therefore of importance for designers to identify the further stages in the preparation of a successful design collection, including the types of equipment which should be at hand.

11.5   THEME BOARDS AND THE PRESENTATION OF IDEAS

As noted in section 11.3, a theme board is often produced in association with a collection of designs. This will often include a selection of the visual sources used in the development of the collection, the intended end use, recognition of the relevant market segment, the colour palette, the raw materials to be used and the means of manufacture.

Textile designers should always show an appreciation of relevant technological innovations, new raw materials and new ways of achieving their aims. Of importance in the creation of any collection of designs are visual development and the simultaneous intention of aiming for an anticipated end use. For a design collection to be successful, it needs first to be manufactured. Manufacturers aim for production within certain financial boundaries. Designs should also be of a type which is perceived as producible from available raw materials and processing technology. Most crucial is that designs should be acceptable to both producers and consumers of the product. This acceptability is best communicated through a theme board. Visits by designers to what are known as trade shows may stimulate the visual content of theme boards. Premier Vision in Paris and New York, Indigo in Paris and Pitti Filati (focused mainly on yarn usage) in Italy all have a role to play.

By the early-twenty-first century, computer-aided design (CAD) and associated computer-aided manufacture (CAM) were integral parts of the textile industry worldwide. The use of digital technology across all design-related activities, from yarn manufacture to dyeing and fabrication, had revolutionised the design expectations of all companies. A website presence, including an e-commerce facility, for each company was the norm. Inclusion of seasonal theme boards on websites was also common.

The anticipated cost of manufacture and the cost to the consumer will vary substantially. It is well known that between the end of textile manufacture and garment retailing (for example), substantial further costs are added. The theme board should give some indication of the types of products anticipated by the designer so that manufacturers can be given some awareness of potential production costs.

11.6   SUMMARY

This chapter recognised the necessity to conduct market research on an anticipated product area at an early stage of the design process. The importance of collages, theme boards and visual observation and representation as means for focusing intentions was stressed. The differentiation between primary and secondary visual research was explained, as was the importance of anticipating an end use at the beginning of the design process. Crucial considerations for the designer should include both the cost of manufacture and the cost to the consumer, and manufacturers should be able to anticipate production costs through consideration of the theme board.

REFERENCES

1. Steed, J., and F. Stevenson (2012), *Textile Design: Sourcing Ideas. Researching Colour, Surface, Structure, Texture and Pattern*, London: AVA Publishing.
2. Steed, J., and F. Stevenson (2020), *Sourcing Ideas for Textile Design*, London: Bloomsbury Visual Arts.

12 Testing and appraisal

12.1 INTRODUCTION

Textile testing is aimed at assessing in advance how a textile may perform in actual use. Designers and manufacturers must select appropriate fibrous raw materials and products from various stages of manufacture, in order to ensure that the production is at acceptable standards; for, if it is not, the future implications for manufacturers can be severe, with the resultant low reputation having a negative impact on future repurchase. To monitor a textile in use prior to launch would be ideal, but frequently this is not possible, though sometimes advance testing is possible. Garments, for example, could be test worn, but this may take several weeks (or even months) to complete. Often, a relatively small selection of products (deemed to be representative) is tested.

Textile designers must therefore ensure that designs produced are suited to the intended end use. A range of practical characteristics must be assessed prior to the application of any design to its anticipated end use. Failure to do so may have detrimental effects on consumer awareness of the collection. Probably the most important practical considerations are colour fastness, tensile strength, stretch and resistance to abrasion, as well as levels of flame retardation, water repellence and absorption, though the range of desirable characteristics is closely dependent on anticipated end use. This chapter identifies the more important physical and chemical tests conducted prior to release of relevant products on the market. Where test results are deemed inadequate, the raw-material input or the system of manufacture must be adjusted.

The catch-all term 'textile testing' covers a multitude of tests aimed at establishing an accurate account of the physical (or mechanical) and chemical properties of textiles at the various stages of processing. Different countries have different established test standards for textile products for sale within their territories, and testing can establish whether these have been complied with. Sometimes testing is done between stages of the processing sequence, often to establish whether machinery has been set correctly or whether concentration of additives needs to be adjusted. Large, well-established, mills will often have in-house testing facilities. In addition, after delivery of fibres, these too should be checked to ensure that quality expectations and agreed specifications are adhered to. Therefore, it appears that raw materials and final products need to be tested, as does each stage of manufacture. Testing may also take place prior to agreement to purchase a consignment of textile products and will be designed to establish whether the products (fibres, yarns or cloths) being purchased are to the standards agreed or required. Textiles intended for one end use will have different requirements from textiles intended for another end use. Textiles destined for use in say ladies' swimwear will have different requirements from cloths destined for use in domestic upholstery, for example; the former may need a resistance to swimming-pool chlorine and saltwater and may also need to have a high light fastness, whereas the latter may need to have good abrasion resistance as well as high flame-retardation characteristics.

Publication relating to the testing of textiles since Booth's monumental work (first published in the 1960s) has been extensive with significant, though often highly specialist, contributions made by Saville (1999), Hu (2004), Westbroek, Piniotalds and Kiekens (2005), Behery (2005), Fan (2005), Pan and Gibson (2006), Hu (2008), Gupta (2008), Hearle and Morton (2008), Schwartz (2008), Miraftab (2009), Bunsel (2009), Chen (2009), Neckâr (2012), Annis (2012) and Veit (2012).

In recent decades there has been interest in the subjective testing or evaluation of textile products. By the late-twentieth century, it was readily apparent that new forms of (manufactured) fibres and finishes provided a wider range of comfort-related properties than was the case fifty years previously (Slater 1997). This progress continued into the early-twenty-first century, to the extent that comfort aspects of cloth, including handle, drape and warmth, played a major role in the marketing outlook of many companies. Relevant reviews and discussions of the more important issues were provided by Das and Alagirusamy (2010), and Song (2011). The objective of this chapter is to identify the key tests carried out in order to assess whether a textile (fibre, yarn or cloth) is suited to the intended end use.

12.2   FIBRE, YARN AND CLOTH IDENTIFICATION

There are numerous tests aimed at testing samples of fibres, slivers, yarns or cloths. It is often the case that it is not desirable to test all the items in a consignment due to time and cost constraints and because many tests are destructive. An important aspect of testing is therefore identification and sampling, where a selection of items from a consignment is made and these are tested in the belief that they are representative of the larger consignment. Extensive discussion of textile statistics and sampling was given by Booth (1968: 8–98) in the well-known treatise *Principles of Textile Testing* (re-published on several occasions since its first publication in 1961). Examination of a cloth's constituent fibres using an optical microscope can be helpful for identification purposes. However, often the expected appearance of fibres can change through applied finishes, for example, during mercerisation. Fibres dyed with dark colours are difficult to examine under a microscope, as light cannot penetrate easily. Also, many manufactured fibres may be difficult to identify as numerous spinneret cross-sections are possible. In cases where optical microscopes can be used, both fibre cross-section and longitudinal views may be necessary. Where fibres have not been dyed, the natural colour should be assessed also. Natural cotton may be white, cream or yellowish; wool may be white, or various browns or blacks; silk may be lustrous white; viscose and many other manufactured fibres may be transparent, white or dullish yellow; carbon fibres may be black. Where fibre identification presents difficulties, various staining techniques and other chemical tests may assist with fibre identification.

The types and characteristics of the fibres used are major considerations as these will determine to a large degree the mechanical performance of the textile. An important consideration is whether the fibre is in a staple or continuous-filament state. If of the former type, the average staple length is of importance as this will determine settings in yarn manufacture. In the case of cotton fibres, the degree of maturity and an estimation of non-fibrous (or trash) content will be of importance, if a collection of fibres is being assessed. In all cases, fibre quality is an important consideration, as well as moisture content, tensile strength, fibre crimp (or waviness) and fineness. As with fibres, yarns may be assessed for numerous characteristics, including a range of physical features such as tensile strength, whether yarns are singles, doubled or plied, amount of twist (turns per inch or per centimetre), the direction of twist, yarn count or yarn hairiness; standard tests and equipment have evolved for all of these, most by the early-twenty-first century including a digital reading of some kind.

Assuming that fibre and yarn content are understood, there are numerous further tests to assess the properties of cloths, including tensile strength, abrasion resistance, drape, cloth cover (area covered by constituent yarns), flammability, handle, and resistance to pilling and shrinkage, as well as thermal properties, cloth thickness, and number of yarns per centimetre or inch in warp or weft directions. Further tests may set out to determine whether certain cloths hold harmful substances such as banned dyes, pesticides, lead or other undesirable heavy metals. Details of relevant tests can be found in numerous textbooks such as Booth (1968) and Saville (1999). It should be noted also that interesting bibliographies focused on each aspect of testing were provided by Booth (1968). While the contents of these were published in the 1950s or before, by the early-twenty-first century many still offered worthwhile sources from which to develop an understanding of the fundamental issues associated with textile testing. A particularly useful treatise covering a wide range of tests was edited by Ahmad, Rasheed, Afzal and Admad (2017). Various important characteristics and tests with reference in particular to cloth drape were reviewed and discussed by Sanad and Cassidy (2016). Some of the more important tests are identified and explained briefly below, but it should be noted that a multitude of tests is possible for fibres, yarns or cloths, though only a small number are mentioned here. It is well known that the foundation of a cloth destined for a particular end use is determined to a large degree by the selection of fibres, the method of yarn manufacture and the means of cloth production.

12.3   COLOUR, STRENGTH, STRETCH AND ABRASION

Colour fastness is the measure of the resistance of the colour in a dyed cloth when exposed to certain agents or conditions. Colour fastness of cloths to light, laundering, chlorinated and non-chlorinated water, and perspiration are crucial considerations depending on the anticipated end use. Numerous tests for colour fastness under different conditions have evolved. Generally, five grades are used in the scale from 5 (signifying no apparent change) to 1 (where substantial change appears to be the case). In the case of fastness to light (where more detail is considered necessary to help with dye selection) a scale of 1 to 8 is used, with 8 representing the highest degree of fastness.

Strength, stretch and abrasion are related closely to fibre content as well as yarn and cloth structures. Appropriate tests are available to assess general physical characteristics (and changes in these) at fibre, yarn and cloth levels.

The tensile strength of a textile is occasionally referred to simply as the breaking strength, which is the maximum recorded force prior to the breaking of a fibre, yarn or cloth (Tubbs and Daniels, 1991: 37). Other related physical characteristics include tear strength and pilling, both of importance at cloth level, but possibly determined by decisions earlier in fibre processing. Tear strength is the force necessary to allow a tear in a cloth to continue once it has begun (Humphries, 2004: 294). Pilling is the formation of globular entangled fibres which may occur during wear or washing. Relevant tests have evolved to assess both tear strength and the pilling resistance. As fibres, yarns and cloths have different levels of elasticity, the stretch characteristics of a cloth will often depend on its constituents as well as its method of manufacture. What is of concern is recovery (or resiliency) after stretching and, as can be anticipated, numerous tests have evolved to assess this. Abrasion resistance relates to the resistance of the cloth to wear, and is of importance in clothing, furnishings and various technical end uses.

12.4   FLAMMABILITY AND ABSORPTION

Flammability is the measure of the tendency of a cloth to burn under specific conditions. Flammable cloths will ignite and will burn. Flameproof cloths will not burn. Flame resistance is the ability of a cloth to resist burning. Flame-resistant cloths will burn but will be extinguished when not in direct contact with flames. Tests have evolved to evaluate cloths for both flammability and flame resistance, often focused on determining the reaction of cloths to heat and fire as well as on the safety aspects of various product categories, such as children's toys, household furnishings and clothing. Some cloths may be thermoplastic and become soft when heated and thus allow heat setting but will melt and drip when heat is intense.

Absorption is the measure of the tendency to take up moisture. A related term is 'wicking', which indicates the possibility of moisture passing along the length of the constituent fibres. These characteristics are of importance in hot climates where perspiration is a major issue. Therefore, linen cloths are particularly suited to hot climates as they absorb water well and wick well.

12.5   SUITABILITY FOR PURPOSE

Textile designers must always have the anticipated end use at the forefront of their minds during the design process. Suitability for purpose is a crucial consideration, and this is determined often by the fibre content and how this has been arranged in cloth format, including how it has been coloured, printed, woven, knitted or manipulated in any other way. Probably the most important driving force is common sense. All essential criteria to ensure that a textile performs well in the intended end use should precede the design process and designers need to identify important performance characteristics at the outset.

Important tests not mentioned above include those that assess the resistance of textiles to ultraviolet rays, crease recovery, flexing, alkalis, acids, perspiration, crimp, lustre, flexibility, drapability, wind resistance, and closely related issues such as seam stretch, slippage, sagging and snagging. Many others have evolved also.

12.6 SUMMARY

It was explained that the phrase 'textile testing' covers numerous tests, designed to ensure an awareness of the performance potential of fibres, yarns or cloths. It was recognised that a fundamental aspect of testing was identification of the content of the textile being tested, and numerous tests had evolved to test not only textiles which were deemed to be ready for some future use, but also newly arrived consignments in advance of further processing, including textiles in loose fibre, sliver, yarn or cloth format. Many tests were destructive, so it was necessary to ensure that samples tested were representative of the wider population of items. The importance of issues such as colour fastness, tensile strength and abrasion resistance (together with many other characteristics) was stressed. A brief review was presented of a small selection of physical and chemical tests.

REFERENCES

Ahmad, S., A. Rasheed, A. Afzal and F. Admad (eds.) (2017), *Advanced Textile Testing Technologies*, London and New York: CRC Press (part of the Taylor and Francis group).

Annis, P. (ed.) (2012), *Understanding and Improving the Durability of Textiles*, Cambridge, UK: Woodhead.

Behery, H. (ed.) (2005), *Effect of Mechanical and Physical Properties on Fabric Hand*, Cambridge, UK: Woodhead.

Booth, J. E. (1968), *Principles of Textile Testing*, third edition, London: Butterworth.

Bunsell, A. R. (ed.) (2009), *Handbook of Tensile Properties of Textile and Technical Fibres*, Cambridge, UK: Woodhead.

Chen, X. (ed.) (2009), *Modelling and Predicting Textile Behaviour*, Cambridge, UK: Woodhead.

Das, A., and R. Alagirusamy (2010), *Science in Clothing Comfort*, Cambridge, UK: Woodhead.

Fan, Q. (ed.) (2005), *Chemical Testing of Textiles*, Cambridge, UK: Woodhead.

Gupta, B. S. (ed.) (2008), *Friction in Textile Materials*, Cambridge, UK: Woodhead.

Hearle, J. W. S., and W. E. Morton (2008), *Physical Properties of Textile Fibres*, fourth edition, Cambridge, UK: Woodhead.

Hu, J. (2004), *Structure and Mechanics of Woven Fabrics*, Cambridge, UK: Woodhead.

Hu, J. (ed.) (2008), *Fabric Testing*, Cambridge, UK: Woodhead.

Humphries, M. (2004), *Fabric Reference*, Upper Saddle River, NJ: Prentice Hall.

Miraftab, M. (ed.) (2009), *Fatigue Failure of Textile Fibres*, Cambridge, UK: Woodhead.

Neckâr, B. (2012), *Theory of Structure and Mechanics of Fibrous Assemblies*, Cambridge, UK: Woodhead.

Pan, N., and P. Gibson (ed.) (2006), *Thermal and Moisture Transport in Fibrous Materials*, Cambridge, UK: Woodhead.

Sanad, R. A., and T. Cassidy (2016), 'Fabric Objective Measurement and Drape'. *Textile Progress*, 47(4):317–406.

Saville, B. P. (1999), *Physical Testing of Textiles*, Cambridge, UK: Woodhead.

Schwartz, P. (2008), *Structure and Mechanics of Textile Fibre Assemblies*, Cambridge, UK: Woodhead.

Slater, K. (1997), 'Subjective Textile Testing'. *Journal of the Textile Institute*, 88(2):79–91.

Song, G. (ed.) (2011), *Improving Comfort in Clothing*, Cambridge, UK: Woodhead.

Tubbs, M. C., and P. N. Daniels (eds.) (1991), *Textile Terms and Definitions*, Manchester: The Textile Institute.

Veit, D. (ed.) (2012), *Simulation in Textile Technology*, Cambridge, UK: Woodhead.

Westbroek, P., G. Priniotalds and P. Kiekens (2005), *Analytical Electrochemistry in Textiles*, Cambridge, UK: Woodhead.

13 Further considerations

13.1 INTRODUCTION

At the end of the second decade of the twenty-first century, the textile industry globally had risen to the challenge offered by digital technology across the full range of manufacturing. The major issue facing society in general was environmental, as the textile industry contributed more than its fair share of environmental pollution. The principal market for textile fibres, by the early-twenty-first century, remained the clothing industry. With processing located mainly in lower-income countries, largely in Asia, clothing manufacture by that time was aimed at producing low-quality, throwaway garments, made available at exceedingly low prices to consumers based in Europe and North America. This cheap, readily available, fashion became known as fast fashion. Through the medium of fast fashion, the textile industry contributed much to a deteriorating environment. By the beginning of the third decade of the twenty-first century, that there was much pressure on manufacturers and retailers to improve the durability of products manufactured. This seemed to offer the best means of safeguarding the future, not only for manufacturers and their workforces but also for consumers. More than any other change or development associated with textile manufacture, fast fashion's contribution to a deteriorating environment has been substantial.

This chapter identifies the nature of late-twentieth- and early-twenty-first-century teaching, research and development relating to textile products and processes; comments on the use of microfibres and nanofibres and on the introduction of environmentally friendly fibres and high-performance textiles of various kinds; outlines aspects of sustainability and ethical manufacture relating to textiles as well as their distribution in clothing form; and reviews briefly the nature of what are known as smart textiles.

13.2 RESEARCH AND DEVELOPMENT

Textile research and development in the late-twentieth and early-twenty-first centuries had focused on new forms of fibre as well as new aspects of fibre, yarn and cloth processing. Among the most important innovations were high-performance fibres, environmentally friendly fibres, extra-fine fibres known as microfibres and nanofibres, digital and 3-D printing, plasma finishing, 3-D clothing production from fibres or yarns, numerous product innovations associated with nonwoven textiles, the introduction of smart textiles and the use of various fibres as additions to products used in high-performance, medical and hygiene areas. The availability of digital innovations across processing was the most obvious feature, with all processing, ordering and transfer digitally controlled. Key participants in the development (and commercialisation) of innovations throughout much of the twentieth and early-twenty-first centuries were fibre and machinery manufacturers, often funded from governmental sources, and based largely in North America, Japan or Europe, as well as research groups in universities. During the late-twentieth and early-twenty-first centuries, technological innovations and their diffusion were therefore of relevance to all areas of textile processing. Also of great importance (by the beginning of the third decade of the twenty-first century) was the greater concern for the environment and ethical manufacture among manufacturers and the expression of a willingness to contribute to the lessening of the carbon footprint associated with textile processing, principally through the use of fibres that were considered to be environmentally friendly and the use of less water in processing. Among retailers, there was the assurance of association only with manufacturers who provided ethical working conditions. A selection of product innovations and a review of apparent environment-related thinking among manufacturers are given further attention below

13.3   EXAMPLES OF PRODUCT INNOVATIONS

Important late-twentieth-century product innovations were microfibres and nano-fibres, both fine manufactured fibres (often of polyester) substantially thinner than a human hair (with nanofibres even finer than microfibres). Often such fibres were included in cleaning cloths for lens and similar surfaces. With yarns that could be knitted or woven, both microfibres and nanofibres were used across a wide range of applications. Light-weight, durable and water-repellent cloths were typical.

In the wake of environmental concerns associated with textile manufacture, lyocell fibres came to the fore. These were fibres manufactured from sources which were regarded as sustainable, and therefore, the fibres themselves were regarded as sustainable. A closed-loop system was employed ensuring that the solvents used in manufacture were recycled, with a recovery rate of 99 per cent (claimed, for example, by the Lenzing Group on their website, in the early-third decade of the twenty-first century).

A fundamental development was the introduction of various high-performance fibres, with enhanced (and often unexpected) properties, able to perform in extremes of one kind or another. Probably the greatest advantage of these new high-performance fibres was their high-strength-to-low-weight ratio, with often outstanding resistance or tolerance to high temperatures, chemical attack, abrasion, cutting and fatigue. In the early-twenty-first century, carbon fibres were probably the most widely used high-performance fibres. With exceedingly high strength and low weight, carbon fibres were used commonly in aviation, aerospace, automobile, military and civil-engineering products. A comprehensive review of high-performance fibres and their range of applications was provided by Hearle (2001), which included contributions dealing with aramid fibres from Rebouillat (2001: 23–61), carbon fibres from Van Dingenen (2001: 156–190), glass fibres from Jones (2001: 191–238), ceramic fibres from Bunsell and Berger (2001: 239–258), chemically resistant fibres from Horrocks and McIntosh (2001: 259–280), and thermally resistant fibres from Horrocks, Eichhorn, Schwaenke, Saville and Thomas (2001: 281–324).

A common end use for many fibres, especially carbon fibres, was within composites where, primarily because of low weight and high strength, certain fibre types replaced the use of aluminium or steel. By the end of the second decade of the twenty-first century, numerous further applications for fibrous structures were apparent in areas such as communications, life support and protection, leisure, sports and automobile manufacture. The phrase 'technical textiles' was used to describe the bulk of these high-performance applications. Horrocks and Anand (2000) provided a good late-twentieth-century review and, more recently, Alagirusamy (2010) gave a review of industrial and medical applications of technical-textile yarns. A brief, though well-focused, review of the use of fibrous raw materials in the automobile sector was presented by Wilson (2019).

13.4   SUSTAINABILITY AND ETHICAL MANUFACTURE

During the late-twentieth and early-twenty-first centuries, there was an awareness that the world's natural resources were being depleted at an accelerated pace, and the impact (or 'carbon footprint') of the textile industry on the physical environment was substantial. The term 'sustainability' was used to refer to the use of the earth's resources to meet the needs of the present without compromising the availability of resources for the future. So the term 'sustainability' entered the textile and fashion vocabulary. It was believed that, if designers and manufacturers sourced fibrous raw materials from textile waste or from pre-used textiles, this would reduce the necessity for newly manufactured fibres, would lower energy consumption and thus help to improve the carbon footprint of the textile industry. In fibre reclamation, cloth was shredded into (slightly weaker and shorter) fibres, used often in mattresses, insulation or as nonwovens of various kinds including cleaning wipes. Textile recycling (another term to enter the fashion vocabulary) was the re-using of fibrous raw materials; however, often fibres were not easily extracted and sometimes the cost of extraction was more than the cost of using fresh, new fibres. It appears to be the case that cheap cotton/polyester or wool/acrylic blends, used typically as constituents of fashion clothing in the early-twenty-first century, were expensive to recycle in any meaningful way. Meanwhile plastic bottles would often be ground and reconstituted into polyester fibres used in the manufacture of fleece-type cloth. Wang (2006) provided a worthwhile review of textile recycling. A further term to enter the fashion vocabulary was 'upcycling', to refer to extending the life of a textile product through possible part-repair and re-using of it in an application not anticipated when it was first produced. Common sense suggests that recycling and upcycling can only, realistically, be considered as temporary contributions towards sustainability, and that a return to the manufacture of durable products would have a greater impact on the positive welfare of consumers, manufacturers and their workforces, and societies in general.

Despite the environmental alarm call outlined above, by the early-twenty-first century it seemed that most manufacturers and retailers, while paying lip service to sustainability and the need to protect the environment, maintained a dominant focus on satisfying the demand for clothing products which were: non-sustainable, fast, unethical and almost throwaway. The phrase 'fast fashion' was used to describe a situation where an inexpensive, though fashionable, garment was produced rapidly under the direction of mass-market retailers (with headquarters largely in North America and Western Europe).

Terms such as fair trade and ethical manufacture were in common use by the early-twenty-first century. These indicated that workers involved at each stage of manufacture were paid a reasonable wage, under clean and safe working conditions, and that child labour was not a feature. By the early-twenty-first century, the term 'organic' was used commonly as well, to indicate that cotton fibres used were produced under agricultural conditions which were regarded as non-detrimental to the environment. In addition, the use of the term indicated that the cotton employed had not been genetically modified, that crop rotation was practised to encourage fertility and, where necessary, only natural means of controlling pests were used. Overseeing numerous fashion collections was the campaign focused on maintaining animal rights, and by the early-twenty-first century the use of real fur had declined, and numerous substitutes were available.

By the end of the second decade of the twenty-first century, societies produced vast quantities of textile and related rubbish, and millions of tons of non-biodegradable waste were generated. Methane and CO_2 emissions, worldwide, were at high and dangerous levels, and it was believed that this led further to global warming and polar-ice-cap deterioration. Recognition of these issues was expressed frequently in a vast range of literature. A small selection includes: Burall (1991), Cooper (1994), Slater (2003), Blackburn (2005), Allwood, McLaursen, de Rodriguez and Bocken (2006), Christie (2007), Fletcher (2008), DEFRA (2009), Armstrong and Le Hew (2011), Tobler-Rohr (2011), Harper (2012), Anon (2015), Mathu (2017) and Kingsley (2019).

By the beginning of the third decade of the twenty-first century, it seems that, in developed economies worldwide, a substantial proportion of clothing purchase was via internet-based providers rather than through what became known as bricks-and-mortar (or B & M) outlets (Matheny, 2019: 7). This led to substantial quantities of packaging materials, often not suited for recycling, ending in landfill. In order to accommodate internet-purchase hesitations relating mainly to colour, texture and fit, return of merchandise by consumers had become simple and cost free; but it seems that abuse of this system resulted in substantial quantities of returns, again adding to landfill (Anon, 2019: 15).

It is recognised that this short section raises many questions and, if anything, only opens the debate yet again. What is certain is that the apparent behaviour of early-twenty-first-century textile manufacturers, clothing retailers and fashion consumers cannot continue into the third decade of the twenty-first century, and substantial re-alignment of resources is required. A changed focus on durability and ethical manufacture in textile, clothing and other product manufacture seems the only route to stimulate a change in demand, on the one hand, and an improved environment in the longer term, on the other.

13.5   SMART TEXTILES

An important development of the early-twenty-first century was electronic textiles of various kinds and the incorporation of these into personal clothing. Often, the term 'smart textiles' was used to refer to such products. It appears that the first mention, in a textile context, to the term 'smart' was in the late 1980s when reference was made to a shape-retaining silk textile (van Langenhove, 2014). Much of the associated background research, innovation and development, as well as the accompanying finance, was from military or health sources with a focus, for example, on responsive camouflage or wound healing, respectively. A well-focused review of possibilities was provided by Berzowska (2012).

Dramatic technological innovations in materials and their use occurred in the latter half of the twentieth century and the early decades of the twenty-first century. Many of these were in the wake of research into high-performance materials associated with outer-space research and the needs of NASA. The discovery of high-performing aramid fibres and polytetrafluoroethylene (commonly referred to as PTFE) are two well-known examples.

Various additions can be made to textiles to ensure responses to climatic or health conditions. These so-called smart textiles (with incorporated sensors of various kinds) are able to respond to changes in temperature, pressure or light, and, by the beginning of the third decade of the twenty-first century, a wide range of devices could be incorporated into clothing textiles and used to monitor both the medicinal and the cosmetic needs of the wearer.

The transition of computing potential from a desktop facility to personal clothing, although a substantial change conceptually, has been helped partially along the way by widespread acceptance of portable smart phones capable of impressive computing possibilities. Indeed, from the outset of the third decade of the twenty-first century, it appeared that the problems associated with wearable computer technology, such as lack of flexibility and low durability of the computing additions, were largely resolved, and that electronic textile innovations were being applied to a wider consumer setting. Important early concerns with smart textiles were the nature of power sources and battery charging, though innovations associated with portable smart phones and the capability of speedy charging from other battery-powered devices have had widespread applications to solve problems thought previously to be insoluble. By the early-twenty-first century, textiles were assuming roles above and beyond what was expected conventionally. Applications were in various areas, including health and medicine, sport and personal physical development, architecture and interiors.

Smart textiles sense and react to their environmental conditions, with different types capable of detecting mechanical, chemical, electrical and thermal, or other sorts of change. These are textiles which can apparently think for themselves and react accordingly, and their development has involved a myriad of expertise beyond conventional textile technology. Smart textiles are used not just in clothing applications but across a wide range of further applications, including automobiles, military uses, aircraft, robotics and medicine, where fibrous structures of various kinds are employed. Only a few examples are identified and discussed in this chapter. Comprehensive reviews of early-twenty-first-century developments were provided by Tao (2015) and Koncar (2016).

13.6   SUMMARY

Numerous technological innovations were of relevance to the global textile industry of the early-twenty-first century. The introduction of microfibres and nanofibres, autolevellers, digital printing, plasma finishing, high-performance fibres and smart textiles are a few examples. The digital revolution had influences throughout the textile and clothing pipeline, not just on manufacture but also across all aspects of commerce and distribution.

Possibly the most important development in textile and clothing manufacture during the early-twenty-first century was the widespread acceptance of fast fashion, which contributed substantially to a deteriorating environment. The production focus of most textile manufacturers worldwide was on non-sustainable, short-lived products, aimed at achieving the lowest-possible manufacturing costs and highest financial turnover. Meanwhile, the related clothing industry in Asia (particularly in Bangladesh) during the early-twenty-first century came under increased scrutiny, following various disastrous events which refocused the attention of Western consumers on ethical manufacture. Nevertheless, even in the aftermath of these disasters and the loss of life, it appeared that the commercial pressures of fast fashion forced textile producers to continue to provide the cheapest possible cloths.

In the early-twenty-first century, as clothing items became relatively cheap, durability declined also, and the manufacture of clothing which had a high impact on the health of the environment became the norm. So, textile processors worldwide were blamed for contributing substantially to a deteriorating environment. But by the early part of the third decade of the twenty-first century, it was suggested in the textile and fashion press that textile and clothing companies had become more firmly focused than before on sustainability, and on using less water and fewer chemicals, and were at least expressing views which suggested concern with the welfare of the workforce. This, of course, would make sound economic sense, but only the passage of time will establish whether these expressed concerns are fruitful and contribute positively to environmental welfare.

REFERENCES

Alagirusamy, R. (ed.) (2010), *Technical Textile Yarns. Industrial and Medical Applications*, Cambridge: Woodhead.

Allwood, J. M., S. E. Laursen, C. Malvido de Rodriguez and N. M. P. Bocken (2006), *Well Dressed? The Present and Future Sustainability of Clothing and Textiles in the United Kingdom*, Cambridge: University of Cambridge Institute for Manufacturing.

Anon (2015), *Clothing Durability Report*, Banbury, UK: Waste and Resources Action Programme (WRAP).

Anon (2019), 'Fashion "Fit" for the Future'. *Textiles* (3):15–17.

Armstrong, C. M., and M. L. A. Le Hew (2011), 'Sustainable Apparel Product Development: In Search of a New Dominant Social Paradigm for the Field Using Sustainable Approaches to Fashion Practice', *The Journal of Design Creative Process & the Fashion Industry* 3(1):29–62.

Bunsell, A. R., and M. H. Berger (2001), 'Ceramic Fibres', in J. W. S. Hearle (ed.), *High-performance Fibres*, Cambridge, UK: Woodhead, pp. 239–258.

Burall, P. (1991), *Green Design*, London: Design Council.

Berzowska, J. (2012), 'Electronic Textiles: Wearable Computers, Reactive Fashion and Soft Computation', in C. Harper (ed.), *Textiles: Critical and Primary Sources*, London and New York: Berg.

Blackburn, R. (ed.) (2005), *Biodegradable and Sustainable Fibres*, Cambridge: Woodhead.

Christie, R. (ed.) (2007), *Environmental Aspects of Textile Dyeing*, Cambridge: Woodhead.

Cooper, T. (1994), 'Beyond Recycling'. *Eco Design*, 3(2).

DEFRA. (2009). *Maximising Reuse and Recycling of UK Clothing and Textiles – Project Summary, Key Findings, Researchers' Recommendations, Methodology & Scope*, London: British Government Papers.

Fletcher, K. (2008), *Sustainable Fashion and Textiles: Design Journeys*, London: Earthscan/ James & James.

Harper, C. (ed.) (2012), *Textiles: Critical and Primary Sources*, 4 volumes (vol. 2 entitled *Production* included reference to sustainability), London and New York: Berg.

Hearle, J. W. S. (ed.) (2001), *High-performance Fibres*, Cambridge, UK: Woodhead.

Horrocks, A. R., and S. C. Anand (eds.) (2000), *Handbook of Technical Textiles*, Cambridge: Woodhead.

Horrocks, A. R., H. Eichhorn, H. Schwaenke, N. Saville and C. Thomas (2001), 'Thermally Resistant Fibres', in J. W. S. Hearle (ed.), *High-performance Fibres*, Cambridge, UK: Woodhead, pp. 281–324.

Horrocks, A. R., and B. McIntosh (2001), 'Chemically Resistant Fibres', in J. W. S. Hearle (ed.), *High-performance Fibres*, Cambridge, UK: Woodhead, pp. 259–280.

Jones, F. R. (2001), 'Glass Fibres', in J. W. S. Hearle (ed.), *High-performance Fibres*, Cambridge, UK: Woodhead, pp. 191–238.

Kingsley, J. (2019), 'Eyes on the Ice'. *National Geographic*, September, pp. 104–115.

Koncar, V. (2016), *Smart Textiles and Their Applications*, Cambridge: Woodhead.

Matheny, R. (2019), 'Slow Fashion Leads to Slow Retail', *Textiles*, (3):6–10.

Mathu, S. S. (2017), *Sustainable Fibres and Textiles*, Cambridge: Woodhead.

Rebouillat, S. (2001), 'Aramid Fibres', in J. W. S. Hearle (ed.), *High-performance Fibres*, Cambridge, UK: Woodhead, pp. 23–61.

Slater, K. (2003), *Environmental Impact of Textiles*, Cambridge: Woodhead.

Tao, X. (ed.) (2015), *Handbook of Smart Textiles*, Singapore: Springer.

Tobler-Rohr, M. I. (2011), *Handbook of Sustainable Textile Production*, Cambridge: Woodhead.

Van Dingenen (2001), 'Carbon Fibres', in J. W. S. Hearle (ed.), *High-performance Fibres*, Cambridge, UK: Woodhead, pp. 156–190.

Van Langenhove, L. (2014), 'Smart Textiles: Past, Present, and Future', in X. Tao (ed.), *Handbook of Smart Textiles*, Singapore: Springer, pp. 1–20.

Wang, Y. (2006), *Recycling of Textiles*, Cambridge: Woodhead.

Wilson, A. (2019), 'Beyond Electric', *Textiles*, (3):22–24.

14 Conclusion

In the drafting of the text for this book during the latter part of the second decade of the twenty-first century, it was obvious that the fundamentals of processing had remained largely unchanged over the few decades up to then. What had changed dramatically was how this processing was organised and controlled. The textile industry, in its best-practice mode, had become highly capital intensive, with digital technology controlling manufacture at each stage. All processing stages, from fibre collection, extraction and assembly, to preparation, yarn manufacture, fabrication, colouration and finishing, were digitally controlled, as was textile testing and all communications relating to buying and sales.

Textile-machinery inventions, when they are patented and realised commercially as fully formed innovations, often require substantial further research and development to make them less restrictive in the type of product produced; in other words, the applicable product range of any textile innovation is only extended through further inventive attention. This, at least, was the case throughout much of the twentieth century, with a possible exception being during war time when physical resources among participants were devoted to other matters. As the years pass beyond the second decade of the twenty-first century, it can be expected that green issues, together with related sustainability, ethical concerns and a focus on durability will receive attention in debates relating to technical development. Stricter environmental regulations worldwide will force manufacturers to consider all aspects of manufacture, and this may well be the economic force necessary for further inventive effort. Recycling will become the norm and all producers will need to focus on the sustainability of their manufacture, as well as the welfare of their employees.

It seems likely that genetic engineering (or possibly some similar procedures) will be focused on the development of fibres with specific physical and chemical properties. Innovative forms of fibres will be manufactured using new chemical combinations or will be created biologically using suitable bacteria. Plasma finishing and digital printing will receive much further research attention and this, in turn, will lead to widespread adoption among processors. By the beginning of the fourth decade of the twenty-first century laser and plasma technologies will offer great potential to the textile industry, with immense possible environmental benefits resulting from reduced use of water and various chemicals, and the lowering of resultant effluent. Further innovations in smart textiles will continue into the mid-twenty-first century and the incorporation of these into clothing and other products can be expected. Further to this, it is expected that nonwoven textiles will continue their widening applicability and new forms will undergo broad acceptance among consumers.

In the longer term, it is anticipated that degrees of automation with digital control over temperatures, speeds and general machine efficiency will increase substantially. The focus of innovation and technological development will remain on lowering labour costs and increasing production speeds. Robotic control in textile manufacture will become common, as will the introduction of whole systems of manufacture concerned simply with the input of fibres and the output of finished cloth, with minimum human intervention in between.

Glossary

This Glossary consists of key terms relating to fibres, yarns and cloths and their processing. Where possible, the emphasis is on raw-material types and products (yarns and cloths). Some terms included may be unfamiliar to textile- and fashion-design students, as well as to specialists in areas other than textiles. The aim is to give a brief definition of the common terms. Trade names and brands are excluded, as it is felt that these are often transitional, with the dominant types often subject to replacement after the passage of only a few decades. At the same time, where possible, an overlap with information given in chapters is kept to a minimum.

The Textile Institute's *Textile Terms and Definitions* (with the online version introduced in the early-twenty-first century) is probably the most comprehensive source of definitions on textile-related matters. The copy used to support the text here is the ninth edition, edited by Tubbs and Daniels (1991). Although some updating has been required, Tubbs and Daniels (1991) provided a good substantial text. The second edition of the *Fabric Glossary* and the third edition of *Fabric Reference*, both by Humphries (2000 and 2004, respectively), were of value when drafting this glossary as well (though, again, substantial updating was required). The reader can refer to the *Guide to Textile Terms*, produced by Anstey and Weston (1997), which is a valuable listing of terms covering fibres, yarns, fabrics, dyeing, printing and finishing. It may be worth referring to Nicholson's (2009) *On Tenterhooks*, which provided a compendium of terms and sayings of textile origin which have entered the wider English language. There are also numerous online compendia, many associated with the websites of professional associations, often providing full-colour illustrations of textile products. Various web-based historical sources, mainly from national-museum websites (such as the Victoria and Albert Museum), are worth referring to as well.

Abrasion:	the wear associated with use.
Absorption:	a term that denotes the willingness of a textile to take up water or other liquids.
Acetate:	a manufactured fibre predominantly of cellulose. The fibre is a close competitor of silk, especially in terms of drape and lustre.
Acid dye:	a dye that has an affinity for protein fibres, especially wool, as well as nylon fibres.
Acrylic:	a manufactured fibre, developed in the mid-twentieth century as a substitute for wool. The fibres are soft, machine washable and with good colour fastness.
Add on:	the dry weight of solids left on a cloth after dyeing and finishing.

Adhesive felt:	a cloth created from a web of fibres bonded together by adhesive material.
Adire:	a resist-dyed cloth associated closely with West Africa.
Adjrak:	a resist- and mordant-dyed, block-printed cotton textile, typically in primary red and blue, with black and white, associated with Sindh in Pakistan and Gujarat in India. The textile, produced in several stages, was most often used as a shoulder mantle, turban, scarf or stole.
Aida:	a woven cotton gauze with an open, even weave, used as a base for embroidery, particularly cross-stitch.
Albert:	a reversible overcoat cloth, often in double weave, with different striped or check designs on the face and reverse sides.
Alginate:	a fibre derived from seaweed sources.
Alpaca:	a fine, soft and shiny fibre from the Alpaca, a camel-like animal native to the Andes.
Angora:	a term used to refer to fibre from the angora rabbit. It is soft and light-weight and has been blended often with wool from sheep.
Aniline:	a dye made from coal tar, which was the first synthetic dye type, introduced in the mid-nineteenth century.
Appliqué:	an embellishment to a fabric's surface, through the addition of separate pieces of fabric.
Aramid:	a synthetic fibre of exceedingly high strength, low weight, good flexibility and impressive heat resistance.
Argyle:	a diamond-shaped knitwear design, often in solid colour, associated with Scotland.
Asbestos:	a family of naturally occurring mineral fibres, renowned for their fireproof properties but associated closely with substantial health risks, both in manufacture and in use.
Autoclave:	a vessel for treating textiles under pressure and/or steam.
Axminster:	associated with a machine-woven carpet with a cut pile. There are four main types (spool, gripper, spool-gripper and chenille).
Baize:	surface-felted (or milled), very durable, smooth and regular woven cloth of wool, processed through a worsted system, used often to cover gaming tables and, in the past, as linings for scientific-instrument boxes and musical-instrument cases.
Balanced weave:	a type of weave where the number and extent of floats in both warp and weft directions is the same for face and reverse sides.

Balbriggan: a soft, napped, jersey-knitted cloth of cotton, wool or a blend, named after a town in Ireland. Typically used as underwear.

Bamboo: a plant fibre that offers numerous advantages over other fibre sources especially cotton. It is exceedingly fast-growing, is found in diverse climatic conditions (from cold mountainous to hot tropical regions) and is widely recognised as offering a sustainable fibre source. It has very high yields, with low water requirements and needs negligible quantities of pesticides or fertilisers.

Bannockburn: a woollen tweed-like cloth in either straight 2/2 twill or herringbone weave, with single and two-ply yarns alternating in both warp and weft directions.

Bar (or barré):   a fault running across the width of a cloth.

Barathea: a term used to refer to either a relatively heavy silk cloth in a broken rib weave or a soft worsted cloth with a pebbled surface, usually in black or dark blue, used for a variety of clothing end uses, especially gent's suitings.

Bark cloth: a nonwoven cloth created by soaking and pounding fibres from the inner bark of trees (often mulberry trees). Various types have been produced worldwide, with the tapa cloths of Tonga, Samoa and Fiji being among the most renowned. In Hawaii, another renowned producer and user, it is known as kapa.

Basic dye: a dye with affinity for acrylic fibres.

Basket weave: a modification of plain weave, with two or more warp and weft threads used in each interlacement rather than one of each.

Bast: fibres such as flax, hemp or jute, obtained from within the stems of plants.

Batik: a method of resist dyeing using wax, rice paste or similar impervious materials to cover part of the cloth during the dyeing process. The parts covered will 'resist' the dyestuff.

Batiste: a soft, light-weight, fine, opaque, plain-woven cloth from flax, cotton, wool or a blend of these.

Batt: a sheet of fibres of regular thickness, but no particular orientation, often used in nonwoven cloth production.

Beam: a flanged roller used to hold warp yarns prior to weaving.

Beaver: a term used to refer to a heavy-weight wool cloth, which has been milled, raised and cut to simulate beaver fur.

Bedford cord:	a cotton cloth with continuous rounded cords or ridges in the warp direction and pronounced lower areas between. Good durability and high strength make it suited as upholstery and as work wear.
Beetling:	a finishing process, traditionally involving hammering dampened linen cloth with wooden hammers.
Beige:	a soft dress cloth in 2/2 wool worsted twill.
Bengaline:	a cloth with a warp-rib appearance, using cotton or wool worsted yarns or with silk warp and worsted weft.
Bias:	a process where a woven cloth is cut at 45 degrees to both warp and weft.
Bi-component:	a fibre manufactured from two polymers.
Binding:	a narrow-woven, knitted or braided cloth added to a more substantial cloth, often to protect an edge or seam.
Biopolishing:	a process where cloths of cellulosic fibres are treated with enzymes to improve the softness of their handle.
Bird's eye:	a term used in both weaving and knitting to refer to a cloth with small spots. In weaving it is considered as a colour-and-weave effect (created on simple woven structures by the systematic insertion of weft yarns of differing colours). In knitting, the cloth is double knit, in a combination of stitches, with tuck stitches creating a hole effect on the cloth's surface, like a bird's eye.
Bison:	a wool-type fibre (not readily available commercially) from the animal of the same name.
Blanket:	a thick cloth, often with a raised finish. The term is used to refer to a thick cloth which acts as a printing base.
Blanket range:	a term used to refer to samples of a woven-cloth collection with colour ways of the same design formed by the insertion of different weft colours.
Blazer:	traditionally, an all-wool flannel cloth, heavily milled, raised and finished with a short nap. Often in solid colour or woven stripes, and used as jacketing cloth.
Bleaching:	the process associated with the removal of colouring matter from a cloth.
Bleeding:	a term often used to refer to the loss or spread of colour during or immediately after the dyeing process.
Blend:	a mixture of different fibres.
Blister:	a weft-knitted cloth with a three-dimensional relief effect.
Block printing:	printing using wooden blocks, hand-carved in relief, with one block per colour. Occasionally, metal blocks may be substituted for the wood equivalent.

Blotch:	a relatively large area of uniform colour in a printed textile.
Bobbin:	a cylindrical holder, with or without flanges, used to hold roving or yarn. The term is used also to refer to a type of hand-made lace in which threads are fed from bobbins.
Bobbin lace:	a type of lace produced by twisting and crossing yarn from bobbins, with yarns held secure by pins on a pillow.
Bolt:	a length or piece of cloth which traditionally varied in dimensions according to the type of fibre-processing industry.
Bombazine:	a term derived from an obsolete French word, applied originally to a lustrous, piece-dyed, twill-woven silk warp and worsted-weft cloth and, later, to cotton. Used commonly as a dress cloth.
Bombyx mori:	the cultivated silkworm.
Bonded:	two cloths combined using a binding agent, often an adhesive of some kind.
Border pattern:	regular edge embellishment found commonly on carpets, shawls or scarves.
Botany wool:	a term used to refer to slivers, yarns or cloths from merino wool.
Bouclé:	a term used to refer to either a fancy yarn twisted from three or four yarns (with irregular curls or loops) or else a mid- to heavy-weight woollen cloth (knitted or woven).
Bourrelet:	a double-jersey weft-knitted cloth, characterised by horizontal ridges on the face.
Box cloth:	a term used to refer to a plain-woven cloth of 100 per cent wool, with firm handle and maximum covering, used often for leggings.
Braid:	a narrow, tubular or flat textile, made from three or more yarns.
Brin:	a term for a single filament of cultivated silk after de-gumming.
Broadcloth:	a term used traditionally to refer to a dark-coloured, twill-woven, heavily milled woollen cloth, though sometimes in other fibres (including cotton or silk). When of cotton, a close, plain-woven, mercerised structure is used, with twice as many two-ply combed-warp yarns as similar weft yarns. In years past, the term applied to cloth that was wider than 70 centimetres.

Brocade: a term used to refer to a non-reversible Jacquard-woven figured cloth, with raised outline, often in medium- to heavy-weight, with figuring (often floral) provided by warp or weft yarns or both, in satin or twill weaves. Often brocades are of cultivated silk or man-made fibres, possibly including metallic yarns.

Brocatelle: a figured furnishing cloth with a pattern in relief caused by warp yarns in satin weave.

Broderie anglaise: embroidery and cut work on a fine, balanced plain-woven base cloth, in bleached flax or cotton, with end uses including hot-weather lingerie, nightwear, blouses, dresses, bedding, curtaining and table linen.

Broken twill: a twill in which the visual continuity is broken through the use of a twill-type weave with an irregular move number.

Brushed: a term used when a cloth has been raised, possibly by being passed over rollers with fine wires.

Brussels: a term used to refer to a carpet with a loop pile woven on a Wilton loom.

Buckram: a term used to refer to a plain-woven linen or cotton, light-weight and loosely woven cloth, impregnated with a stiffener or filler of some kind, used as inter-face, providing support for cuffs and collars, and as interior reinforcement for wallets, handbags and similar items.

Buckskin: a closely set, milled and cut woven cloth from merino wool. Similar to but heavier than doeskin cloth.

Buffalo cloth: fairly coarse-woven wool from carded and dyed yarns, with a heavily napped surface, originally intended as blankets but used also in apparel as coatings in heavier weights and as jackets and shirting in lighter weights.

Bulk: a term which refers to the volume or space taken up by a yarn.

Burl: a small lump of entangled fibres, slub or other impurity found in woven wool cloth.

Burling: removal of slubs or other flaws from the reverse side of the cloth.

Butcher cloth: a closely woven, medium- to heavy-weight cloth of the type used in the past for butchers' aprons. Often from 100 per cent flax yarns, the cloth was often in plain weave with slubs in both warp and weft.

Cable: where two or more yarns are twisted together. Alternatively, in knitting, where a vertically oriented cabled effect is created. Cable-knitted cloth is double knit, made with loop transfer. A rope-like appearance is a characteristic feature.

Cake: a cylindrical holder of continuous-filament yarn, often viscose.

Calendering: the finish achieved by passing cloth between two rollers to give a smooth and lustrous result.

Calico: a balanced, plain-woven cloth, of a light-beige colour and rough texture, of mid-weight, often from 100 per cent cotton, unbleached and undyed, and intended as a print surface.

Cambric: a fine, light-weight, closely woven cloth in plain weave, often in linen or cotton, sometimes with stiffener added, used as handkerchiefs.

Camel: specialty hair fibres from an animal of the same name.

Camlet: a fine, lustrous, plain-woven cloth of silk, specialty hair or wool used especially for suiting.

Campbell twill: a twill cloth of fine wool or worsted, also known as Mayo twill.

Canvas: a term used to refer to a plain-woven cloth, in cotton, linen, hemp or jute, often with a close-packed warp and an association with firmness and strength. Related terms include awning, duck, sail and tarpaulin cloths.

Carbon fibres: manufactured fibres, predominantly of carbon.

Carbonisation: the removal, using acid, of non-fibrous matter from wool or specialty hair fibres.

Carpet: a textile floor covering, often woven or occasionally felted.

Casein: a protein fibre made from constituents in milk.

Casement: a light- to medium-weight, closely packed, woven cotton, using thick warp yarns. Used as curtaining, table linen or upholstery.

Cashgora: a specialty hair fibre obtained from the cashgora goat (bred from a male angora goat and a female cashmere goat with the intention of obtaining fibre with attributes from both animals).

Cashmere: a specialty hair fibre obtained from the undercoat of the *Capra hircus laniger* or Asiatic goat.

Catalyst: an added substance, often used in textile finishing, which aids a chemical reaction of some kind, yet takes no part in that reaction.

Cavalry twill: a firmly woven steep-twill cloth, often in wool (carded and sometimes combed), used traditionally for cavalry trousers and, in more modern times, as coatings, suits and uniforms.

Centre stitching: in the case of a double-woven cloth, this is where a series of stitching yarns lying between two cloths is interlaced occasionally with each cloth during the weaving process and thus holds the two cloths together.

Challis: a light-weight, balanced, though relatively open, plain-woven worsted dress cloth, originally in fine cashmere and silk, often used as a printing base.

Chambray: a smooth, light-weight, plain-woven cotton cloth, traditionally of linen, often with a dyed warp and white weft yarns, intended often for women's or children's wear.

Charmeuse: a woven, light-weight silk or manufactured filament cloth, in satin weave, used for ladies' lingerie and evening gowns.

Checked: a term used to refer to a cloth with two series of parallel stripes oriented at 90 degrees to each other. Often with an underlying 2/2 woven twill in wool. Examples include various tartans, shepherd's check, hound's-tooth check, gun-club check, buffalo check, guard's check, window-pane check, tattersall check, glen check, pin check and Prince of Wales check.

Cheese: a cylindrical, flangeless, yarn holder.

Cheese cloth: a light-weight, sheer, open-structured, plain-woven cloth usually made from carded cotton yarns, used originally for wrapping food, especially cheeses. Used also for ladies' and children's wear and for general drapery. Due to open structure, the cloth requires little ironing.

Chenille: a term used to refer to a yarn with a cut pile (secured by two yarns twisted together). Chenille cloth, often used in upholstery, contains chenille yarns (a type of fancy or effect yarn).

Cheviot: a term used traditionally to refer to cloth manufactured from wool sourced from a breed of sheep originally from the Cheviot Hills (Scotland). More commonly, the term is used to refer to a heavy-weight, rugged tweed from various wool sources with a harsh handle.

Chiengora: a fibre from dog's fur, processed as if wool from sheep. Not readily available commercially.

Chiffon: traditionally, a very light-weight, sheer, plain-woven, silk cloth, often with alternate 'S'- and 'Z'-twisted crêpe yarns in warp and weft directions respectively which imparted a slight puckering and some stretch. Used principally in ladies' evening gowns.

China grass: another term used to refer to ramie (a bast fibre).

Chino: high-quality, strong, lustrous, combed cotton with a steep twill in mercerised cotton, probably pre-shrunk. Often used in hot weather, in trousers or military uniforms. Originally dyed khaki colour.

Chintz: a glazed, balanced, plain-woven (in carded yarns), light-weight, cotton cloth, often printed in large-scale floral patterns or dyed in solid colour. Used mainly for curtaining and occasionally in outerwear and aprons. Cloths may be half glazed or fully glazed, through friction calendering only in the former case and by coating the cloth in starch prior to friction calendering in the latter case.

Chrome dye: a form of dye with an affinity for wool.

Ciré: a woven or knitted cloth impregnated with a wax, passed between calender rollers to impart a waxy appearance.

Clip spots: a technique where the floats of warp or weft yarn are cut to expose a spotted design.

Clouded yarn: a type of yarn in which two component yarns of different colours alternatively cover or hide each other along the yarn's length.

Cloqué: a woven double cloth with a blistered effect resulting from the use of yarns of different constituent fibres or twist (with different responses to finishing).

Cocoon: the casing of silk used to protect the chrysalis within.

Coir: a fibre extracted from the fruit of the coconut tree.

Colour-and-weave effect: the effect obtained when a simple woven structure (such as plain weave or 2/2 twill) is combined in a systematic way with differently coloured weft yarns. Resultant examples include dogstooth check, birdseye or Prince of Wales check.

Combed: a term generally associated with fine, regular worsted-spun yarns, which have undergone a combing process.

Compacting: finishing processes that lower the possibility of excessive shrinkage during laundering.

Composite: a term used often to refer to a product that combines a solid of some kind with fibrous material.

Compound cloth: a term used often to refer to a woven cloth consisting of combined layers, with each layer having its own warp and weft yarns.

Cone: a conical-shaped yarn holder.

Cop: a yarn holder possibly from a ring-spinning frame.

Cordage: a twisted yarn structure where components have been twisted or braided with the intention of sustaining heavy loads.

Corduroy: a sturdy cloth, often of cotton, with cords or ridges. Higher qualities use combed and mercerised yarns. Often, after weaving, one side is coated with glue (later removed) to hold the pile in place prior to cutting. There are various views on the origin of the word, but the derivation from the French phrase *corde du roi* is probably the most common. One set of warp and weft yarns is used to weave the foundation and an extra set of weft yarns is used to create vertical rows of pile floats. End uses include: trousers, jackets, work wear and occasionally home furnishings.

Corkscrew: a warp-faced weave, with a steep-twill feature.

Cortex: The inner part of many animal-sourced fibres, seen when viewed under a strong microscope.

Cotton: the most-used cellulosic fibre.

Count: an expression of the length per unit weight or weight per unit length of a yarn or filament; these are referred to as indirect or direct-count systems, respectively.

Counting or piece glass: a mounted magnifying glass with a base which measures in either inches or centimetres, so the observer can readily work out the density of warp or weft yarns or loops in a cloth and can more readily see the interlacement of yarns.

Course: a row of loops across the width of a knitted cloth.

Cover: a term used to refer to the extent to which an area of a cloth is covered by its component yarns; a cloth considered to be tightly woven with a relatively high number of warp and weft yarns per centimetre or per inch would be deemed to have a high cover factor whereas a woven net-type cloth (with a relatively small number of warp and weft yarns per centimetre or per inch) would be deemed to have a low cover factor.

Crabbing: a stabilisation process for woven worsted cloth, involving hot water being added to the cloth with warp-ways tension and the subsequent cooling and drying under continued tension.

Crease resistance: the ability of a cloth to resist or to recover from creasing during use. Crease recovery is the further related measure.

Creel: a frame for holding yarn pirns, cones, cheeses or other supply packages during processing.

Crêpe: a term used to refer to numerous textile types, most commonly as a type of woven textile characterised by a puckered or wrinkled face, possibly due to the use of high-twisted 'S' and 'Z' yarns, the use of a particular weave, or by the induction of differential shrinkage in the finished cloth. Used for ladies' wear, linings and home furnishings.

Crêpe de chine: a cloth often of low-twist warp yarns and high-twist weft yarns, in plain-woven silk or continuous-filament manufactured fibres. Used largely in ladies' wear.

Cretonne: a printed cotton furnishing cloth heavier than chintz.

Crewel cloth: an embroidered cloth associated with Kashmir (India).

Crimp: a term used to refer to the waviness of a fibre or yarn.

Crinoline: a stiff, light-weight, plain-woven cloth often made from cotton yarn in warp and horsehair yarn in weft. The cloth is used often as a support for hems.

Crocking: a term used to refer to the fastness of dyes to rubbing or similar abrasion.

Crushed velvet: a cloth with a pile in various orientations after finishing.

Cuprammonium (or simply, 'cupro'): a regenerated cellulosic fibre.

Damask: Jacquard-woven figured linen cloth, often consisting of satin and sateen weaves, possibly named after the city of Damascus (Syria). Common uses include upholstery, curtaining and cushions as well as table linen.

Damasquette: a damask with more than one weft yarn type (often of a different colour).

Degumming: removal of sericin (silk gum) prior to further processing.

Delaine: a printed, plain-woven, light-weight wool cloth.

Delustering: a process that can reduce the sheen of man-made fibres and yarns.

Denim: a rugged, warp-faced 2/1 or 3/1 twill-woven cloth, of carded cotton yarns, indigo-blue dyed in the warp and undyed in the weft. Used as clothing, often as trousers or jackets. The term may have derived from a twill sailcloth associated with Nimes (in France).

Dent: one reed space between adjacent reed wires.

Derby: a knitted cloth in which all the loops of six adjacent wales are intermeshed in one direction and all the loops of the next three wales are intermeshed in the opposite direction, and so on.

Devoré (or burn-out technique): a technique in which one or more components of a multi-component cloth is dissolved, removed or etched by the addition of a chemical of some kind which leaves the remaining components unaffected. Often the devoré addition is printed so that only portions of the sensitive components (warp and/or weft yarns) are removed.

Diced weave: produced from woven structures with repeats split into quarters and the component warp and weft actions

	reversed, giving opposite directional effects in each quarter.
Dimity:	a plain-woven cotton cloth, characterised by vertical ribs, cords or stripes. Often used for summer wear.
Direct dye:	a dye with affinity for cellulosic fibres.
Discharge:	a term used in printing, to refer to the removal of dye by chemical means (often simply referred to as a discharge paste). Sometimes the discharge paste may include a colour not affected and may therefore re-colour the discharged area.
Disperse dye:	a water-insoluble dye with an affinity for cellulose acetate and other hydrophobic fibres.
District checks:	a general term used to refer to woollen cloths with checked embellishments often in bold contrasting colours; associated often with different regions.
Distressed:	a technique in which a cloth is made during finishing to appear aged.
Dobby:	a mechanism attached to a loom aimed at controlling the rising and lowering of heald shafts, often to produce relatively small geometric figuring.
Doctor:	a metallic blade used to remove unwanted material such as print paste during processing.
Doeskin:	a lightly raised, medium-weight cloth, tightly woven in wool (sometimes merino), often in a twill weave.
Dogtooth:	a check wool cloth also known as hound's tooth.
Donegal tweed:	a plain-woven wool textile characterised by a random distribution of brightly coloured slubs or flecks within the yarns, produced traditionally in County Donegal (north-west Ireland).
Double atlas:	a warp-knitted cloth with two sets of yarns making diagonal movements in opposite directions.
Double cloth:	a term applied to a woven cloth consisting of two cloth components held together by self-stitching, centre-stitching or interlacement arrangements.
Double jersey:	a term applied to a range of knitted cloths created on a rib or interlock basis.
Double knit:	a weft-knit cloth, commonly in polyester or wool, which uses interlock stitches and its variations, involving two pairs of needles set at an angle to each other. The structure is stable and compact and does not curl or unravel.
Double piqué:	rib-based weft-knitted cloth.
Doubling:	a term used often to refer to the combination of slivers which, when drafted (or attenuated), show enhanced regularity.

Doupion (or dupion):   a term used to refer to a double silk cocoon, with resultant yarn and cloth with similar names.

Drape: a term used to refer to how a cloth hangs when held at one or more points.

Drill: a term used to refer to a durable woven twill cotton cloth, usually piece dyed for uniforms, work wear and tents.

Duchesse: a traditional Belgian lace design consisting of floral and leaf components, using relatively heavy yarns. The term is used also to refer to a closely woven, fine satin cloth with long warp floats.

Duck: a tight, plain-woven tough and durable heavy-weight cloth, often of cotton or flax yarns, like canvas.

Duffel: a heavy, low-grade, wool cloth, often used for short, hooded coats known as duffel coats.

Dungaree: a 3/1 or 2/1 twill cloth, used for working overalls, like denim, but with the lower qualities being piece dyed.

Dupion: originally a silk from doupion (or dupion) yarns.

Durability: a term used in the context of cloths which perform well in their intended end use over a long time duration.

Edging: a term used to refer to a narrow cloth designed to be attached to the edge of a larger cloth.

Elastane: a manufactured fibre which can be extended to up to three times its original length and can, subsequently, recover to its original unstretched length.

Elastomeric: a term used to refer to a yarn with high extensibility and recovery.

Elongation: the extent to which a sample extends in length during a tensile test.

Elysian: a soft woollen cloth with extra surface weft yarns.

Embossing: the production of a design on a textile, probably through the use of an engraved, heated metal cylinder known as a calender against a relatively soft cylinder of cotton or compressed paper.

Embroidery: embellishment to the face of a cloth through hand or machine needlework.

Emerising: a suede-like finish, produced by passing cloth between emery-covered rollers.

Empress cloth: a twill-faced woollen dress cloth.

End: a term used in weaving to refer to a single warp yarn.

Eskimo: a piece-dyed, all wool, overcoating, with a 5-end satin face and twill on reverse side.

Estamemes: a loosely woven, 2/2 twill, worsted dress cloth with a rough finish.

Face finish: a technique in which the face of a cloth is finished differently from the reverse side.

Faconné: Jacquard-woven cloth often in a single colour with small scattered motifs.

Fancy yarn: a yarn with various effects along its length. Varieties include bouclé, chenille, cloud, spiral, fleck, gimp, knickerbocker, loop, snarl, knop, stripe, eccentric and slub.

Fearnought: a thick, heavy wool cloth used traditionally to cover port holes or as a covering for powder magazines.

Fellmongering: the process of pulling wool from sheep's skin.

Felt: a web of entangled or matted sheep's wool or specialty hair fibres, consolidated by mechanical agitation, heat and moisture.

Flannel: a soft, light- to medium-weight, plain- or twill-woven cloth, generally in wool (though sometimes in cotton), with slight felting on the surface, often brushed, both front and back (known as double napped), with a characteristic soft handle. Checked (tartan-type) clothing, shirts, blankets, bed sheets and pyjamas are typical end uses.

Flannelette: a type of cloth often made from carded cotton yarns and plain woven (with lower twist in weft rather than warp), napped (or raised) on one or both surfaces. Used as shirting, sportswear and children's wear.

Flax: a natural, cellulosic, bast fibre from which linen is manufactured.

Fleece: a type of fabric sometimes weft-knitted in carded cotton fibre, possibly in sliver form, and lightly raised, or at other times, twill-woven in soft-twisted long fibres of wool. Fleece wool is shorn from living sheep.

Flocking: a technique which produces a cloth with a suede-like surface created by attaching short fibres to the surface through adhesion or electrostatic charge.

Florentine: heavy woven cotton cloth, often in 3/1 twill used mainly for overalls or uniforms.

Flushing: heavy woollen coating made in Flushing (Holland).

Folded (doubled or plied): a yarn which comprises two or more singles yarns twisted together, maybe described as two-ply, three-ply, etc.

Foulard: a light-weight, soft, twill-woven cloth, in silk or silk blends, often with small-scale print designs.

Friction calendering: a technique in which a cloth is passed between two moving rollers, one highly polished, engraved and heated, rotating faster than the other softer roller, with a resultant glazed effect.

Friezé:	a heavy wool cloth, milled and raised with nap in one direction hiding the underlying weave; used as overcoating.
Full cardigan:	a knitted cloth, produced on two sets of needles, with lengthways ridges on both face and reverse sides.
Fully fashioned:	a term used to refer to the shaping of a panel of a garment during the knitting process.
Fustian:	a heavy-weight, hard-wearing, woven cloth, often with linen warp and cotton weft.
Gabardine:	a cloth often of fine, combed wool (though sometimes in cotton), in warp-faced 2/1 twill, with high-twist warp yarns, producing a clearly visible diagonal effect. Commonly used in suiting, trousers and raincoats.
Galloon:	a narrow ribbon used as a band for men's hats or as binding for ladies' court shoes.
Garnett:	a machine like a card but with rollers covered with metal teeth rather than pins; used to reclaim waste fibres.
Gassed:	a term used to refer to the singeing of yarn or cloth to remove unwanted surface fibres.
Gauze:	often loose, plain-woven light-weight cloth, in carded cotton yarns. Not durable, but may be used as curtaining, and often as medicinal bandages. In gauze weaving the term refers only to cloth constructed on leno principles
Georgette:	a strong, light-weight, sheer-woven cloth, heavier than chiffon, of highly twisted silk or other fine filament yarn, in balanced plain weave.
Gingham:	often a check cloth, of dyed cotton yarns (combed in the high-quality examples), in a balanced plain weave, of medium weight, used in garments and as table linen.
Glass:	an exceedingly strong continuous-filament fibre but with very low flexibility.
Glazing:	a finishing technique, in which a cloth is given a smooth, glossy surface, often through applied heat and/or pressure.
Granite:	a crêpe weave, with short floats scattered over the surface, with a repeat of between sixteen and twenty weft yarns.
Grecian:	a weave based on a counter-change principle, with floats of warp and weft yarns often producing a cellular effect on both the face and reverse sides of the cloth.
Grey (or greige):	unfinished cloth often used where aesthetics are not a consideration.

Grosgrain:	a medium-grade firm cloth with weft-ways rib in plain weave, with fine warp yarns more numerous and finer than weft yarns.
Habit:	a high-quality woollen cloth, generally dyed in dark shades with a napped finish; used as a cloth in ladies' riding costumes.
Haircord:	a plain-woven cotton cloth with alternate coarse and fine rib lines in a warp direction.
Half-cardigan:	a knitted cloth, regarded as a variation of full cardigan, but on the face the spaces between the lengthways ridges are closer. The reverse side of each has a similar series of ridges.
Half Milano:	a weft-knitted, double-jersey rib cloth.
Handkerchief linen:	a light-weight linen close-woven in balanced plain weave, used for blouses, shirts and dresses.
Handle:	the feel (or hand) of a cloth which may, for example, be coarse, fine, warm, cool, slippery, limp or crisp.
Harris tweed:	a coarse wool cloth with yarns from a blend of dyed fibres, associated with the Isle of Lewis in the Outer Hebrides (Scotland). Other tweed names derive often from their place of manufacture (e.g. Cheviot tweed, Bannockburn tweed and Donegal tweed). Typically, Scottish tweeds are produced in a right-hand twill weave, whereas some Irish tweeds (such as Donegal tweed) are produced in plain weave. Often, yarns are carded only, so are rough and hairy.
Hemp:	a bast fibre used commonly in ropes and twine.
Henrietta:	fine, lustrous, twill cloth with fine silk warp and fine botany-wool weft.
Herringbone:	a broken-twill cloth, usually of wool, with a repeating central line, and reversed twill, or zigzag effect, on either side resembling the backbone of a fish.
Hessian:	a rough, uneven, loose, balanced plain-woven jute cloth (sometimes in hemp).
Holland:	traditionally a fine, plain-woven, medium-weight linen or cotton cloth with beetled or glazed finish, made in many countries in continental Europe, especially Holland; used mainly for interlinings.
Home spun:	a term applied to hand-woven twill cloth of tweed character with yarns hand-spun from local domestic sheep's wool.
Honeycomb:	a woven cloth, with the warp and weft yarns forming ridges and hollows, with cellular appearance.

Hopsack:	often of hemp or jute, used as sacking to hold hops, in very loose plain or basket-type weave, with good wrinkle resistance.
Hound's tooth:	a cloth, of varying weights, characteristically with an irregular check design, often in 2/2 twill using black and white yarns. Known also as dogtooth or gun-club check.
Huckaback:	an absorbent cotton or linen cloth often used for towelling.
Hudson's Bay blanket:	a heavily milled and raised coarse wool cloth first supplied to the Hudson's Bay Company in 1780.
Ikat:	a technique of binding and dyeing warp and/or weft yarns in predetermined areas prior to weaving. Yarns are bound with a dye-resistant material (banana leaf traditionally in Indonesia, and often polypropylene in the modern era).
Imperial sateen:	a cloth with densely packed weft in eight-end sateen weave, smooth or raised.
Indigo:	a natural blue dye from the leaves of certain plants (especially *Indigofera tentoria*, a plant native to tropical zones, particularly in South Asia) used traditionally to dye cotton. Regarded as the oldest dye employed for textile use.
Inkle:	an ancient term for a narrow cloth.
Intarsia:	a patterned single-knit cloth, often using multi-coloured yarns, with no floats and identical appearance on face and reverse. Typically used in shirts, blouses and sweaters.
Interlacing:	in the case of a double cloth, the point at which some of the yarns (warp and/or weft) from one cloth simply interchange with the yarns from the other cloth during the weaving process.
Interlining:	cloth placed between the lining and the outer layer of clothing to help shape the garment; it offers a wide variety of cloth possibilities.
Interlock:	a double-knitted cloth often of fine-cotton or (occasionally) filament yarns, produced on a circular machine. The cloth does not curl at the edges when relaxed.
Irish poplin:	a hand-woven poplin from silk warp and 3-fold worsted weft, with silk on face covering the weft and a moiré finish.
Italian:	a 5-end sateen lustrous cotton cloth used mainly as a garment lining.

Jaconet: a light-weight, plain-woven, lawn or muslin-type smooth cloth with a slightly stiff finish.

Jacquard knit: a single cloth, made using a circular knitting machine with a Jacquard mechanism. Offers numerous combinations of colours, textures and floats. The word was used initially to refer to the selection mechanism associated with weaving and thus also refers to highly patterned woven cloths.

Jaffer: a plain-woven cotton cloth with warp and weft yarns in different colours.

Jappe: a plain-woven cloth of square construction originally in silk yarns.

Jean: a 2/1 warp-twill-faced cotton cloth used mainly for overalls and casual wear.

Jersey: a term used to refer to numerous cloth types, some relatively light-weight and some much heavier. Inserted yarns may give embellishment, stability, cover or comfort. The cloth is single-weft knit in plain stitch showing vertical lines on the face and horizontal lines on the reverse side, with good stretch properties horizontally but bad stretch properties vertically; it curls at the edges when released from tension. The name is probably derived from the island of Jersey, and the characteristic hand-knitted fishermen's sweaters.

Jute: a coarse, tan-coloured, bast fibre, produced chiefly in South Asia.

Kalamkari: a cloth from eighteenth-century (and before) India, involving repeated dyeing, using resists, mordants and hand painting.

Kapok: a brittle vegetable fibre, suited best as lifejacket stuffing due to its high buoyancy.

Kashmir silk: a term applied to plain-woven silk cloth, embroidered or printed, often with characteristic Kashmiri (or Paisley) motifs.

Kashmiri shawls: in wool from the cashmere goat, often in interlocking-twill tapestry weave, or sometimes embroidered, featuring the characteristic motif.

Kernmantel: a braided rope used in rock climbing.

Kersey: a firm, compact, lustrous woollen cloth, diagonally ribbed or twilled, heavily milled and finished with a short nap; similar to melton.

Khadi: a term applied to a wide variety of cloths, of hand-spun and hand-woven cotton (occasionally blended with other fibres). Renowned for its simplicity.

Khaki: a light-weight, soft, twill cloth, from cotton, wool or blends, often used in jackets and skirts, as well as in military and police uniforms.

Knit-de-knit: a type of crimp yarn, from unravelled knitted cloth which has been printed or heat set.

Lahore: piece-dyed cashmere cloth with small-scale dobby effects.

Lamb's wool: fibre from young sheep.

Lamé: a term used often to refer to numerous woven-cloth types with a metallic effect (achieved most frequently using fine metallic ribbons around core yarns, but sometimes by spraying a metallic resin).

Lampas: multi-coloured figured drapery or upholstery cloth, similar to brocade, made from silk, cotton, rayon or combinations, with two sets of warp yarns.

Lawn: a light-weight, crisp, woven cloth made from fine (mainly combed) cotton or linen, with a fine, even thread count. Cotton lawn may be bleached white, printed or dyed in the piece. Used often as handkerchief cloth.

Lawn finish: a light-starch applied to lawn or other fine cloths to give a crisp finish.

Left-hand twill: a term used to record the direction of a twill line in a woven cloth. A left-hand twill shows a twill line passing from lower right to upper left.

Leno: a woven (gauze-type) strong and stable cloth produced through moving alternate warp yarns from side to side.

Linen: a cloth made from flax yarns, though the term has gained wider applicability to refer to cotton cloths as well.

Linoleum: a term given to floor covering with woven jute-cloth backing.

Linsey-woolsey: a rough twill or plain-woven cloth, with flax or cotton warp yarns and wool weft yarns.

Lisle: highly twisted, two-ply, high-quality cotton yarn used in hosiery.

Llama: a specialty hair fibre from an animal of same name.

London shrinking: a finish specific to wool cloths.

Loom state: cloth from the loom without further processing.

Loop pile: an uncut cloth pile.

Loop transfer: process of moving loops in a knitted structure with the intention of shaping or fashioning.

Lustre: the surface sheen of a fibre, yarn or cloth.

Lyocell: an environmentally friendly regenerated cellulosic fibre.

Madapolam: a bleached or plain dyed cotton with a soft finish used for ladies' wear.

Madras: a light-weight, 100 per cent carded cotton cloth, yarn dyed and hand-woven in balanced plain weave with a check design.

Madras net cloth: an open, gauze-type, cotton curtaining cloth with patterning through use of extra weft.

Marionette: a mechanism which controls the movement of shuttles in narrow-cloth weaving.

Marl yarn: a term used to refer to a two-coloured yarn.

Matelasse: a double cloth with quilted appearance, commonly with two sets of warp yarns and two sets of weft yarns (arranged 2 face and 1 back in both warp and weft directions).

Melton: a warm, windproof, heavy-weight cloth, known commonly as blazer or duffel cloth (though the latter of these is generally of heavier weight). Woven in a right-hand twill from 100 per cent wool or wool blends, using carded wool, the cloth is dyed in the piece, fulled (felted) and sheared closely to give a surface of little lustre but with good resistance to soiling and general wear. Used mainly for over-coatings and school uniforms.

Meltonette: a light-weight cloth, otherwise resembling melton cloth, used predominately for ladies' wear.

Mercerisation: a process of treating cotton by immersing yarn or cloth in a solution of caustic soda, which swells the fibres and improves dye affinity, strength and lustre.

Merino: soft, fine, high-quality wool from the merino sheep.

Metallic: a fibre or yarn made predominantly from metal (often aluminium), used often to embellish a cloth's surface.

Microfibres: exceedingly fine manufactured fibres.

Milano rib: a weft-knitted, double-jersey, rib cloth.

Mixed weft: a defect caused by using weft yarn from different lots, each of which may appear the same in the yarn holder but may lead to weft-ways optical errors in cloth format.

Mock leno: a woven cloth with the surface appearance of an open, square mesh.

Modacrylic: a general name for manufactured fibre with a greater resistance to combustion and various chemicals than standard acrylic.

Modal: a regenerated cellulosic fibre.

Mohair: silky specialty hair fibres from the angora goat.

Moiré: an optical illusion which can be obtained when a woven cloth of continuous filament is wet. The effect can be made durable when cloth containing thermoplastic fibres is heated. Often the term is used to refer to watered silk.

Moleskin: a heavy, durable, raised-cotton cloth, in imitation of mole fur, often used as work clothing, with lighter weights employed in various fashion end uses.

Moquette: a firmly woven, cut or uncut, warp-pile cloth, used often in public-transport upholstery.

Mousseline: a fine, plain-woven, transparent cloth in silk, cotton or wool.

Mungo: a wool obtained from rags or other waste-wool textiles.

Muslin: an inexpensive light-weight, loose, plain-woven cotton cloth, sometimes used in the grey and at other times bleached and dyed. The cloth's name may have been derived from the city of Mosul (in northern Iraq).

Needle cord: thinly ribbed cotton corduroy.

Needle-punched felt: a nonwoven cloth produced using a series of barbed needles, which penetrate and entangle fibres presented in a web.

Nep: entangled fibres forming a small knot, found on cloths or yarns.

Net: an open-mesh cloth with a firm structure (knitted, woven or knotted).

Nettle: regarded as the source of a highly sustainable bast fibre, and for this reason may be adopted increasingly through the twenty-first century.

Nylon: a generic name for a class of manufactured fibres, known for its high strength, and flexibility, with excellent abrasion resistance.

Organdie (or organdy): a balanced plain-woven 100 per cent cotton cloth.

Organza: a sheer and strong cloth in balanced plain weave, using high-twist and plied-silk yarn (though manufactured fibres such as nylon or polyester have been used in the early-twenty-first century).

Ottoman: a heavy cloth with weft-ways ribs.

Oxford: a 100 per cent cotton cloth usually in a rib or basket weave, with regular fine warp yarns and thicker, soft-twisted weft yarns. When warp is dyed and weft left undyed, the cloth is known as Oxford chambray. Often used as shirting, dresses, sportswear or pyjamas.

Paisley: a category of design motifs, associated with the town of Paisley in Scotland, based on motifs originating on hand-woven shawls from the province of Kashmir in India.

Panama: a plain-woven, light-weight cloth used for tropical suitings.

Patola: a double-ikat cotton or silk cloth produced in India and reputedly imitated in Indonesia.

Peach: a finish achieved through abrasion or some chemical addition to a cloth.

Percale cloth: a soft, crisp, plain-woven dyed or printed cloth, often of high-quality cotton, used as bed covers.

Piece dyed: cloth dyed as a full length, after it has been woven, knitted or produced otherwise.

Pigment: a substantially insoluble colouring matter used often in printing, where it attaches to the surface of the fibre.

Pile: cloth characterised by tufts or loops of fibres, or yarns, standing up from the surface, possibly with a similar effect on the reverse side. Examples include velvets, corduroy, terry towelling and chenille cloth, as well as hand-knotted or tufted carpets.

Pilling: when fibres entangle, during laundering or use, forming collections of fibres in small balls attached to a cloth (easily detached if of staple fibres, but with difficulty if of continuous filaments).

Pina: a leaf fibre from the pineapple plant.

Piqué: a durable cloth with weft-ways rounded cords and pronounced spaces between. Wadding (or reinforcing) weft yarns are often used to help pronounce the weft-ways cords.

Plain: a term used to refer to a single-jersey, weft-knitted cloth, with a different appearance on front and back.

Plain weave: the simplest, but most stable, woven structure, where each warp yarn raises over one weft yarn and under the next, taking up a checker-board-type configuration.

Pointelle: a double-knitted cloth, which looks like lace (with holes from transferred stitches). Used for ladies' and children's wear.

Polyacrylonitrile: a manufactured fibre known commonly as acrylic.

Polyamide: a manufactured fibre known commonly as nylon.

Polyester: a manufactured fibre of high strength (though lower than nylon), with good abrasion resistance and low absorption of water (so dries quickly).

Polyethylene: a manufactured polyolefin fibre.

Polyolefin: a manufactured fibre type (which includes both polyethylene and polypropylene fibres).

Polypropylene: a manufactured polyolefin fibre.

Polyvinyl chloride:	a manufactured chlorofibre.
Pongee:	a fine, soft silk cloth.
Poplin:	a firm, plain-woven cloth originally from silk, or silk and wool. More commonly, from the early-twentieth century onwards, in cotton and cotton blends, with carded or combed yarns (depending on quality). Used for shirts, dresses, pyjamas, raincoats, trousers and sportswear.
Prince of Wales check:	a colour-and-weave effect, in grey, white and black with fine red lines for definition. Used often for suitings.
PTFE:	polytetrafluoroethylene, used in many forms including as a manufactured fibre.
Punto-di-Roma:	a weft-knitted, double-jersey cloth.
Purl knit:	a cloth which appears the same on face and reverse. Alternate knit and purl stitches in one wale are the feature. With the cloth being identical on both sides, the reversible nature is a feature. Does not curl and is more stretchable in a lengthways direction. Often used in bulky sweaters, but speed of production is very slow.
Raffia:	a vegetable fibre extracted from the leaves of the raffia palm.
Ramie:	a lustrous bast fibre known also as China grass.
Rayon:	a manufactured fibre regenerated from cellulosic sources such as wood pulp.
Reactive dye:	a class of dye used on cellulosic and protein fibres.
Reed:	the comb-like device at the front of a loom through which warp yarns are spaced.
Rep:	a plain-woven cloth with weft-ways ribs.
Rib knit:	a single-weft-knit cloth, with the loops of some wales drawn to the cloth back while others are pulled to the cloth face. Due to the underlying structure, rib knit has good insulation properties as well as crossways stretch, and is used where close fit (as in sleeves, socks and waist bands) is desired.
Right-hand twill:	where the twill line runs from the lower left to upper right. Also known as a 'Z' twill.
Rope:	a multi-ply thread, of great strength, traditionally made from hemp or jute.
Rubber:	a naturally occurring substance, famously highly elastic, but with low strength.
Rug:	a term used to refer often to a small, hand-woven carpet.

S-twisted: a term which refers to fibres in a yarn (when viewed if held in a vertical position) that take up the formation of the central line of the letter 'S', oriented from upper left to lower right.

Sailcloth: a heavy, strong, closely woven cloth of flax, cotton, polyester, nylon or aramid, used for sails on boats.

Sandblasting: a technique used to acquire a faded look, when a cloth is exposed to a blast of aluminium oxide (which appears like sand).

Sanforising: a commercial process aimed at pre-shrinking cloths prior to their use in garments.

Sateen: a weft-faced weave, often with more softer-twisted weft yarns, though may be of mercerised cotton, plain or printed. The small number of interlacement points are arranged to avoid optical lines and to produce a smooth cloth. Often confused structurally with satin (which, by contrast, has a dominance of warp yarns on the cloth's face).

Satin weave: has a dominance of warp yarns on the face, often consisting of cultivated silk or possibly manufactured continuous-filament fibres. Often interlacing on a 4/1 basis, with each warp yarn floating over four weft yarns. The small number of interlacement points are arranged to avoid optical lines and to produce a smooth, lustrous cloth.

Saxony: a fine, soft, woollen cloth.

Scrim: a light-weight, open-weave, coarse cloth. The term is used also to refer to a base used for the production of laminated or coated cloth.

Sea Island: a type of cotton, which offers the longest and finest fibres.

Seersucker: a lightly puckered cloth, most commonly produced in 100 per cent carded cotton in balanced plain weave. The puckering effect may be created by shrinkage in parts after the application of caustic soda; otherwise a similar effect is created by sets of yarns given different tensions from adjacent sets.

Self-stitching: in the case of a double-woven cloth, this is where yarns from one cloth occasionally interlace with yarns from the other cloth.

Semi-worsted: a process which includes some of the stages typical of the worsted spinning system.

Serge: a woven cloth often consisting of wool (in the form of worsted cloth), though possibly silk or continuous-filament manufactured fibres, in a 2/2 right-hand twill, with a twill line visible on both face and back. Often used for suiting, the cloth is commonly piece dyed in a navy blue.

Shantung: a plain-woven rough silk cloth, with a general sheen. Slub yarns with rib effects in the weft-ways direction are a feature. Used in bridal gowns and dresses.

Sharkskin: a smooth durable cloth of wool.

Sheeting: either a closely woven (in plain or twill weave) cloth employed for apparel end uses or a cloth with various household uses (including bed linen). Often of cotton, the apparel variety, when coarser, is known also as muslin, a balanced, close-woven, plain-weave cloth, again in cotton.

Shepherd's check: a term associated with shepherds in Scotland wearing lengths of cloth, from undyed black and white wool.

Shetland: a light-weight wool cloth with lightly twisted yarns.

Shoddy: a low-grade wool cloth from fibre obtained from rags or similar waste.

Silk: fibre in either continuous-filament form or in staple form from a cultivated moth or wild moth, respectively.

Singe: a process in which surface fibres on a cloth are removed through burning.

Single jersey: a plain, weft-knitted cloth created on one set of needles.

Single piqué: a double-jersey, weft-knitted cloth created from both knitted and tuck loops.

Sisal fibre: a strong, flexible, leaf fibre used for ropes, twine and coarse cloth.

Sizing: the application of a gelatinous substance to warp yarns prior to cloth construction, making the yarns stronger and more pliable, with the objective of protecting them from abrasion during cloth formation. Removal of sizing (a process known as de-sizing) takes place after cloth formation and prior to any further processing.

Slub: a part of a yarn, which is irregular in diameter. Often considered a flaw but may have been purposely created for aesthetic reasons.

Space dyed: the creation of multi-coloured effects in yarns at intervals along their length.

Stitch: another term for a knitted loop.

Sulphur dye: a class of dye used on cellulosic fibres.

Swatch: a small sample of a cloth.

Taffeta: a crisp, soft, smooth lustrous cloth often of silk, though occasionally in continuous-filament manufactured fibres. Close-woven in plain weave, with up to twice as many warp as weft yarns, the cloth is smooth, but with fine weft-ways ribs. Ribbed or wavy characteristics with occasional moiré effects are a feature. When differently coloured yarns are used in warp and weft, the cloth is referred to as shot taffeta or shot silk.

Tartan: a light-weight, plain-woven wool cloth woven in a multitude of differently coloured warp- and weft-ways stripes to produce a characteristic check. The association of particular tartans with specific clans or families in Scotland is a relatively recent development.

Teasle: a dried seed head (known traditionally as Fuller's thistle) used to raise a pile on a wool cloth.

Terry: often a carded cotton yarn cloth, woven with a warp-pile (of uncut loops) on both sides for better absorption. Where loops are cut, the cloth is known as velour. Classification is through weight, pile structure and finishing. Similar cloths can be knitted (using a raschel warp-knitting machine) to produce strong cloth with good absorption and soft handle at low cost; often cotton yarns are used as the pile.

Thermoplastic: a term used to refer to manufactured fibres that can be permanently fused/shaped at high temperature.

Ticking: a general term which refers to cloth used to cover mattresses or pillows.

Tissue: a richly coloured woven silk (or manufactured fibre) with characteristic metallic yarns. Often used in ladies' wear.

Triacetate: a manufactured fibre, less absorbent than acetate, but can be machine washed and has good wrinkle recovery compared to acetate.

Tri-axial weaving: the interlacement of two warp yarns with one set of weft yarns in such a way that the three sets form equilateral triangles. Such cloths have excellent bursting, tearing and abrasion resistance.

Tricot: a warp-knitted cloth, often in continuous-filament yarns of nylon or polyester, with possible additions such as spandex. Produced on a tricot machine, the resultant cloth is stable and not prone to unravelling. Varieties include simplex and soufflé.

Tuft: part of a pile of a tufted carpet, velvet or corduroy, formed by a length of yarn.

Tulle: a net cloth with hexagonal openings. In weaving this is a fine, plain-woven net of silk yarns.

Tussah: a variety of wild silk obtained from the hatched cocoons of various wild moths, so in staple-fibre form, and often with a natural pale-brown colour.

Tweed: a roughly woven heavy-duty wool cloth, often in 2/1 twill.

Twill: a term used to refer to woven cloth with a clearly defined diagonal effect (which can be of varied steepness and can orientate upward or downward from left to right). Twills involve the creation of warp-ways floats (or weft-ways floats on the reverse).

Twist: the process of making adjacent fibres more compact and into close association with each other, through turning a collection of fibres known as a sliver. Twist is measured by the number of turns per unit length and by the direction of twist, which is deemed to be either 'Z' or 'S' when the yarn is viewed when held both lengthways and vertically.

Twist-on-twist: a term used when a doubled yarn is twisted with the same direction of twist as the single-yarn components. The result is often a twist-lively structure prone to snarling.

Union: a cloth combining flax-warp and cotton-weft yarns, often in plain weave.

Vat dye: a class of dye used on cellulosic fibres.

Velour: a woven or knitted cloth with a surface pile.

Velvet: a soft-pile cloth, available in numerous weights and qualities. Often, better qualities are of silk. The pile is created from extra warp yarns. When complete, the cloth may be cut or uncut. Sometimes, velvet is produced as a double cloth with a pile in between which is then cut to form two cloths. Velvet may be embossed or sculptured, and may include a crush-proof finish. End uses include trousers, dresses, jackets and other clothing, as well as bed covers and curtaining.

Velveteen: a cotton cloth, which, before cutting, is similar to corduroy, but with a weft pile across the full width of the cloth and not in parallel rows as with corduroy.

Vicuna: a fine, specialty hair fibre from an animal of the same name.

Viscose: a type of regenerated cellulosic fibre from wood pulp, cotton linters (extra-short fibres) or other sources.

Voile: is a cloth originally of flax or sometimes silk but generally of cotton in the modern era. The best-quality cotton types are in balanced open plain weave, with combed fibre. A crisp cloth using highly twisted (often two-ply, twist-on twist) yarns is characteristic. The

	cloth is often gassed (or singed) to remove protruding fibres. The result is a light, sheer cloth used for nightwear, lingerie, shirts and dresses.
Wadding:	a layer of fibres used as padding.
Warping:	a process which aligns warp yarns in parallel in readiness for use in textile construction.
Warp printing:	a process where a sheet of warp yarns is printed prior to weaving, with a resultant effect like warp ikat.
Web:	a sheet or layer of fibres of regular thickness, as delivered from the carding engine.
Webbing:	a narrow-woven cloth, with the function of bearing heavy weights.
Whipcord:	a woven cloth of worsted-spun wool, in a steep, pronounced right-hand twill and a clear finish.
Yak:	a specialty hair fibre from the animal of the same name.
Z-twist:	where the fibres in a yarn (when viewed vertically) take up the same direction as the central line in the letter 'Z'.

REFERENCES

Anstey, H., and T. Weston (1997), *Guide to Textile Terms*, London: Weston Publishing Limited.

Humphries, M. (2000), *Fabric Glossary*, second edition, Upper Saddle River, NJ: Prentice Hall.

Humphries, M. (2004), *Fabric Reference*, third edition, Upper Saddle River, NJ: Prentice Hall.

Nicholson, E. (2009). *On Tenterhooks*, Bradford: Wool Press.

Tubbs, M. C., and P. N. Daniels (1991), *Textile Terms and Definitions*, Manchester: The Textile Institute.

Index

Z

Printed in the United States
By Bookmasters